Math Notebook: Multiplying and Dividing Whole Numbers, Decimals, & Fractions

Copyright © 2024 by Lataejha Borden

ISBN: 979-8-9892856-2-4

For more information about the copyright policy or to request reprint permissions email service@closelycaptured.com

Contents

Multiplying Whole Numbers

1 - 25

Multiplying Decimals

27 - 39

Multiplying Fractions

41 - 50

Dividing Whole Numbers

52 - 69

Dividing Decimals

71 - 83

Multiplying Fractions

85 - 92

Anchor Charts/Graphic Organizers Reproducible

93 - 108

Glossary

108 - 113

"There is more than one way to solve a problem" –Math Teacher

This book is dedicated to all the people determined to help young mathematicians conquer challenges.

Introduction

Learning multiplication and division facts and strategies is an essential skill for elementary students. As parents, tutors, and educational instructors you play a crucial role in supporting your scholars mathematical development. This guide aims to provide you with a step-by-step approach and visual aids to help teach and reinforce multiplication and division strategies effectively.

By following this guide and utilizing the visual aids, you can effectively teach how to solve multiplication and division problems by using various strategies with your scholar. Remember to be patient, provide encouragement, and make learning enjoyable. With consistent practice, your scholar will gain confidence and mastery in multiplying and dividing, helping them be successful in their mathematical journey.

This book has strategies and visual aids to help your learner understand mathematical concepts.

Multiplication
of
Whole Numbers

Multiplication of Whole Numbers

What I Know

What I want to Know

Multiplication Concept

Introduce your child to the concept of multiplication using a place value chart.

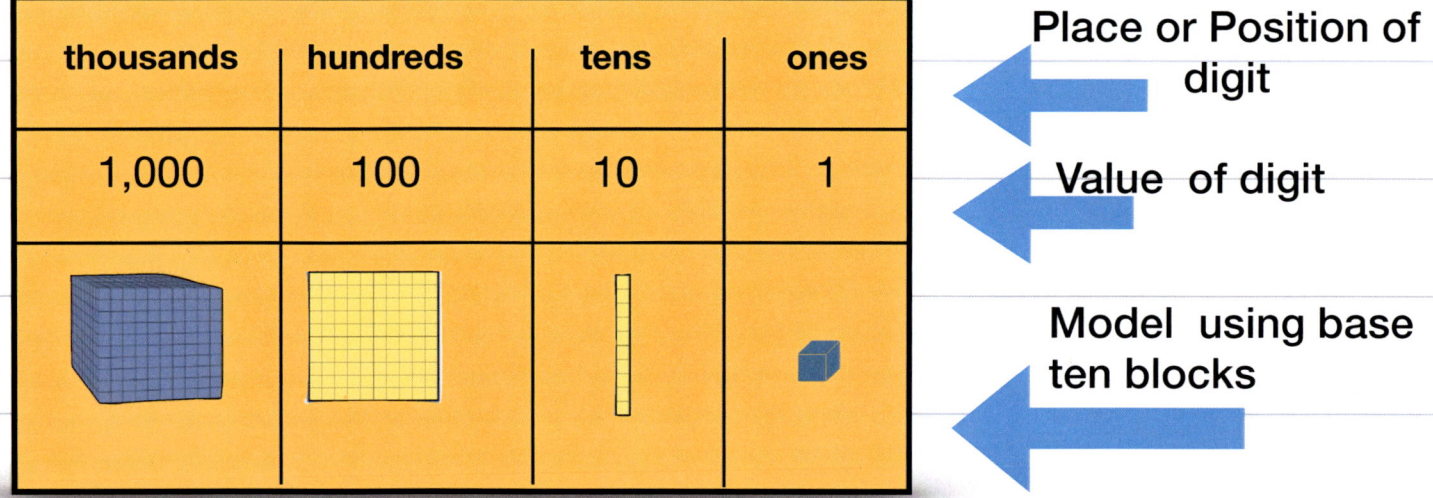

thousands	hundreds	tens	ones
1,000	100	10	1

Place or Position of digit

Value of digit

Model using base ten blocks

The place/position determines the value of the **digit**. Base ten blocks help to model how the value of digits increase.

Whole Number Place Value

As we move from one place to the next the value increases by 10. When we shift from the ones to the tens the value increases by 10. When we shift from the tens to the hundreds the value increases by 10. When we shift from the hundreds place to the thousands place the value increases again by 10.

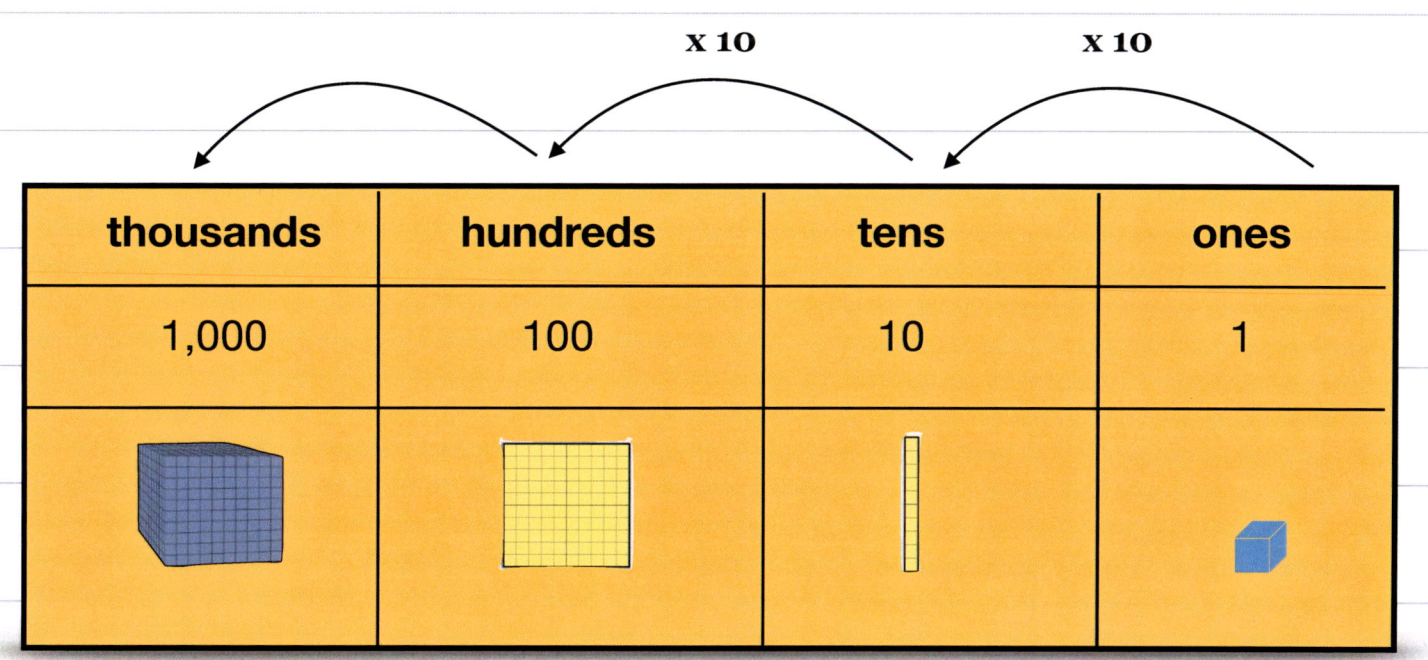

Multiplication Fluency

Introduce your child to the concept of times tables, which are a set of multiplication facts for specific numbers (e.g., 2 times table, 3 times table). Mathematicians can practice learning their basic facts by working on fluency daily, using multiplication charts

Work on one number at a time. Fill in the blanks completely going down and then across before moving to the next number in the order

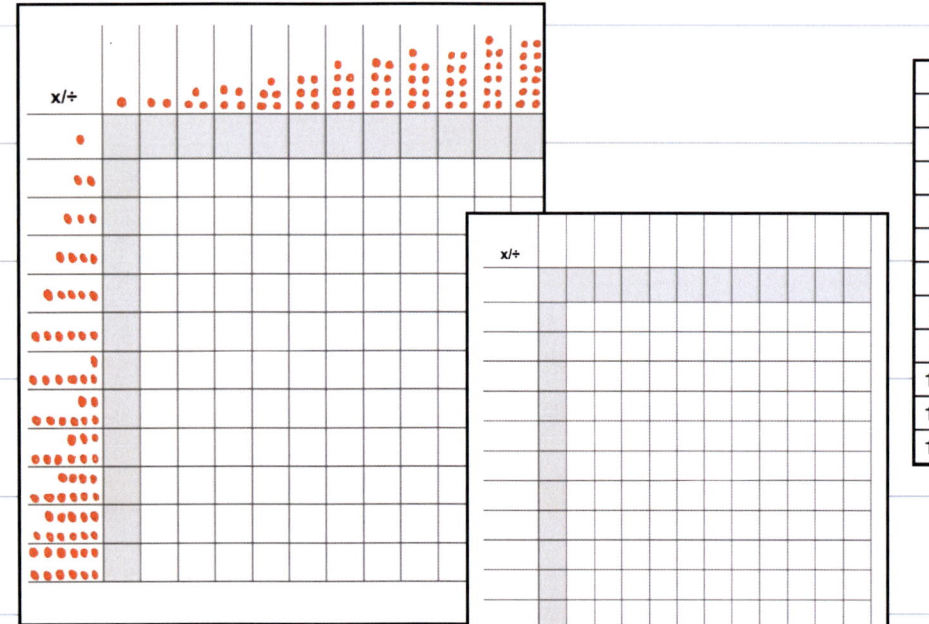

1	2	3	4	5	6	7	8	9	10	11	12
2	4	6	8	10	12	14	16	18	20	22	24
3	6	9	12	15	18	21	24	27	30	33	36
4	8	12	16	20	24	28	32	36	40	44	48
5	10	15	20	25	30	35	40	45	50	55	60
6	12	18	24	30	36	42	48	54	60	66	72
7	14	21	28	35	42	49	56	63	70	77	84
8	16	24	32	40	48	56	64	72	80	88	96
9	18	27	36	45	54	63	72	81	90	99	108
10	20	30	40	50	60	70	80	90	100	110	120
11	22	33	44	55	66	77	88	99	110	121	132
12	24	36	48	60	72	84	96	108	120	132	144

To build speed and fluency of facts, complete the chart in this order
1, 2, 5, 10, 11
4, 8, 12
3, 6, 9
7

Multiplying using a Multiplication Chart

ex. 8 x 7

1. Move across the top row to the multiplicand

2. Move down the blue shaded area to the multiplier

3. Where the multiplicand and the multiplier meet is the product.

1	2	3	4	5	6	7	8	9	10	11	12
2	4	6	8	10	12	14	16	18	20	22	24
3	6	9	12	15	18	21	24	27	30	33	36
4	8	12	16	20	24	28	32	36	40	44	48
5	10	15	20	25	30	35	40	45	50	55	60
6	12	18	24	30	36	42	48	54	60	66	72
7	14	21	28	35	42	49	56	63	70	77	84
8	16	24	32	40	48	56	64	72	80	88	96
9	18	27	36	45	54	63	72	81	90	99	108
10	20	30	40	50	60	70	80	90	100	110	120
11	22	33	44	55	66	77	88	99	110	121	132
12	24	36	48	60	72	84	96	108	120	132	144

8 x 7 = 56

Multiplication Basics

$$3 \times 5 = 15$$

Product the answer

Multiplicand - What is being multiplied

Multiplier - How many times to multiply

Think about 3 times five as five groups of three

Commutative Property of Multiplication states that changing the order does not change the product.
So, think about three times five's three groups of five

$$5 \times 3 = 5 \times 3$$

Think about adding a number repeatedly. For example, 3 x 5 means adding 3 five times: 3 + 3 + 3 + 3 + 3 or adding 5 three times: 5 + 5 + 5

Multiplying using Equal Groups/Equal Sets

Imagine you brought 5 shirts and each shirt cost $3. How much did the shirts cost in all?

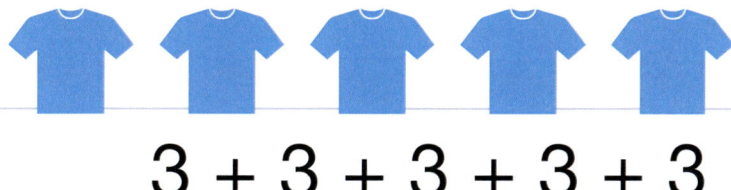

$$3 + 3 + 3 + 3 + 3$$

three groups of five

$$5 \times 3 = 15$$

The total cost of the shirt is $15

Your school bake sale sold 5 boxes of cookies. Each box contained 5 cookies. How many cookies were sold in all?

$$5 + 5 + 5 + 5 + 5$$

five groups of five

$$5 \times 5 = 25$$

There are a total of 25 cookies

Multiplying using Arrays

three times four

3 x 4

columns   rows

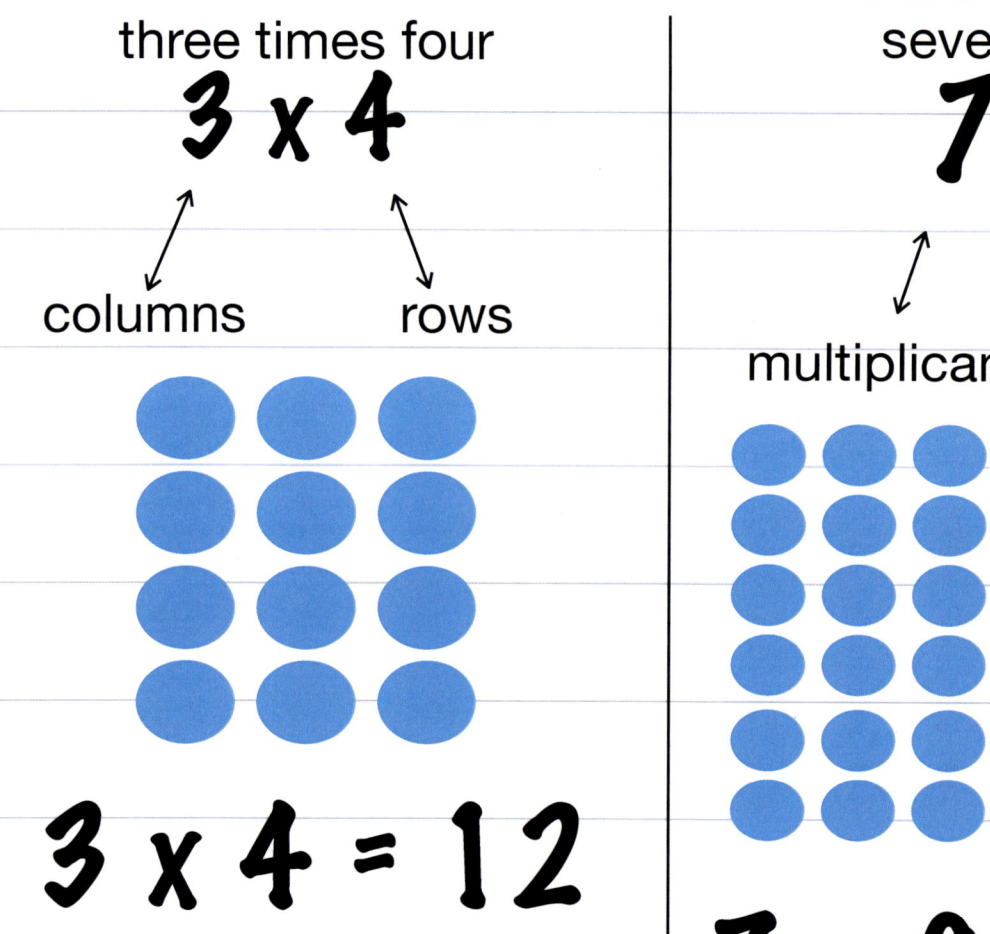

3 x 4 = 12

seven times six

7 x 6

multiplicand   multiplier

7 x 6 = 42

Multiply using Multiplication Patterns

When there is one or more zeros at the end of a **factor** it can help you see a pattern

We know 3 x 5 = 15

3 x 50 = 150
3 x 500 = 1,500
3 x 5,000 = 15,000
3 x 50,000 = 15,000
3 x 500,000 = 1,500,000

What pattern do you notice?

First multiply the basic facts which is 3 x 5
Think about quantity of zero(s) in the factor
The same quantity of zeros(s) should be in the product

When there are zero(s) in both factors

30 x 40 = 1,200
300 X 40 = 12,000
300 x 400 = 120,000
30 x 4,000 = 120,000
300 x 4,000 = 1, 200,000

First multiply the basic facts which is 3 x 4
Think about quantity of zero(s) in each factor
The same quantity of zeros(s) should be in the product

Multiply by Multiples of 10, 100, & 1000

A **multiple** is the **product** result of one number multiplied by another number.

$$4 \times 5 = 20$$

multiple

$$4 \times 50 = 200$$
$$40 \times 500 = 20000$$
$$400 \times 500 = 200000$$
$$4000 \times 500 = 2000000$$

Multiply using Break Apart Strategy
Distributive Property

The **distributive property** states you can break apart factors to equal the sum of two or more **addends**.

$$4 \times 6 = 2 \times 6 = 12$$
$$+ 2 \times 6 = 12$$
$$12 + 12 = 24$$

To distribute means to divide or give a share or part of something.

$$46 \times 3 = 20 \times 3 = 60$$
$$+ 20 \times 3 = 60$$
$$+ 6 \times 3 = 18$$
$$60 + 60 + 18 = 138$$

*an **addend** is a digit added to another

Multiply Whole Numbers using Area Model/Box Model Strategy

Using area model is a visual strategy students can use to solve multi digit problems

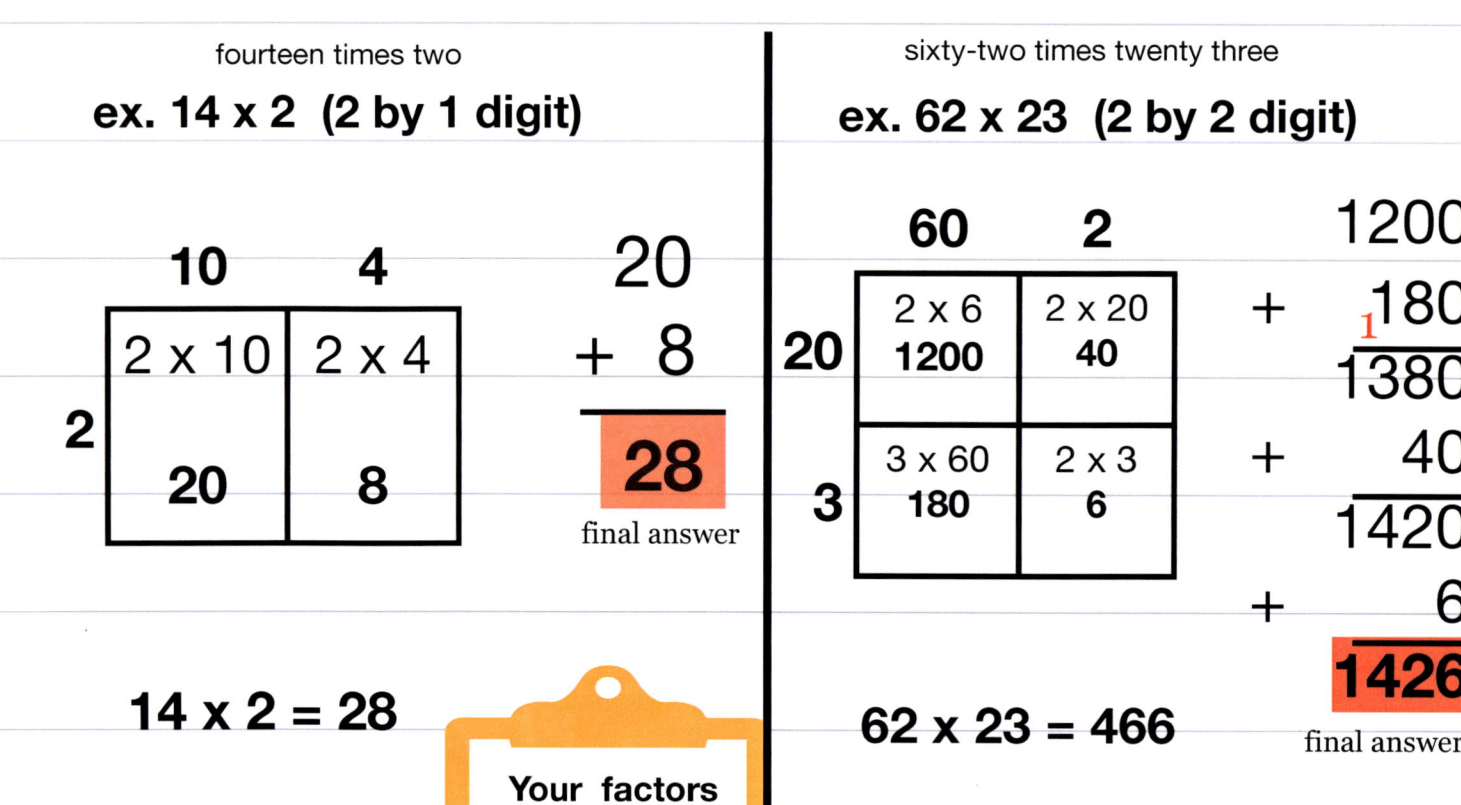

fourteen times two

ex. 14 x 2 (2 by 1 digit)

sixty-two times twenty three

ex. 62 x 23 (2 by 2 digit)

14 x 2 = 28

62 x 23 = 466

Your factors tell you how many rows and columns to create.

Multiply Whole Numbers using Area Model/Box Model Strategy

one hundred forty-three times two

ex. 143 x 2 (3 by 1 digit)

	100	**40**	**3**
2	2 x 100	2 x 40	2 x 3
	200	**80**	**6**

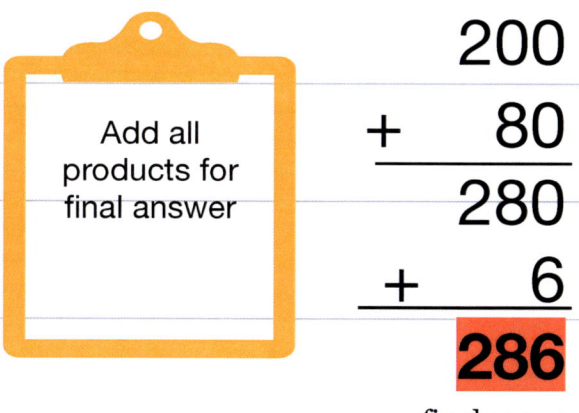

Add all products for final answer

$$
\begin{array}{r}
200 \\
+\quad 80 \\
\hline
280 \\
+\quad\quad 6 \\
\hline
\mathbf{286} \\
\end{array}
$$

final answer

143 x 2 = 286

one hundred forty -three times twenty-three

ex. 143 x 23 (3 by 2 digit)

	100	**40**	**3**
20	20 x 1000 **2000**	20 x 40 **800**	20 x 3 **60**
3	3 x 100 **300**	3 x 40 **120**	3 x 3 **9**

$$
\begin{array}{r}
2000 \\
+\quad 800 \\
\hline
2800 \\
+\quad 300 \\
\hline
3100 \\
+\quad 120 \\
\hline
3220 \\
+\quad 60 \\
\hline
3280 \\
+\quad\quad 9 \\
\hline
\mathbf{3289} \\
\end{array}
$$

143 x 23 = 3289

final answer

Multiply Whole Numbers using Area Model/Box Model

one hundred sixty-two times two hundred thirty one

ex. 162 x 231 (3 by 3 digit)

	100	60	2
200	200 x 100 **20000**	200 x 60 **12000**	200 x 2 **400**
30	30 x 100 **3000**	30 x 60 **1800**	30 x 2 **60**
1	1 x 100 **100**	1 x 6 **60**	1 x 2 **2**

When you find the product of each factor, we must add.

$$
\begin{array}{r}
20000 \\
+\ 12000 \\
\hline
32000 \\
+\quad 400 \\
\hline
32400 \\
+\ 3000 \\
\hline
35400 \\
+\ 1800 \\
\hline
37200 \\
+\quad\ 60 \\
\hline
37260 \\
+\quad 162 \\
\hline
37422 \\
\end{array}
$$

final answer

162 x 231 = 37422

final answer

Remember your factors tell you how many rows and columns to create.

Let's Practice: Multiplying Whole Numbers using Area Models

twenty four times fifty nine

24 x 59

forty five times seventy three

45 x 73

Remember to break apart the factors. Think about tens and ones

Let's Practice: Multiplying Whole Numbers using Area Models

one hundred sixty-six times seventy one

166 x 71

Remember to break apart the factors. Think about hundreds, tens and ones

two hundred sixteen times sixty three

216 x 63

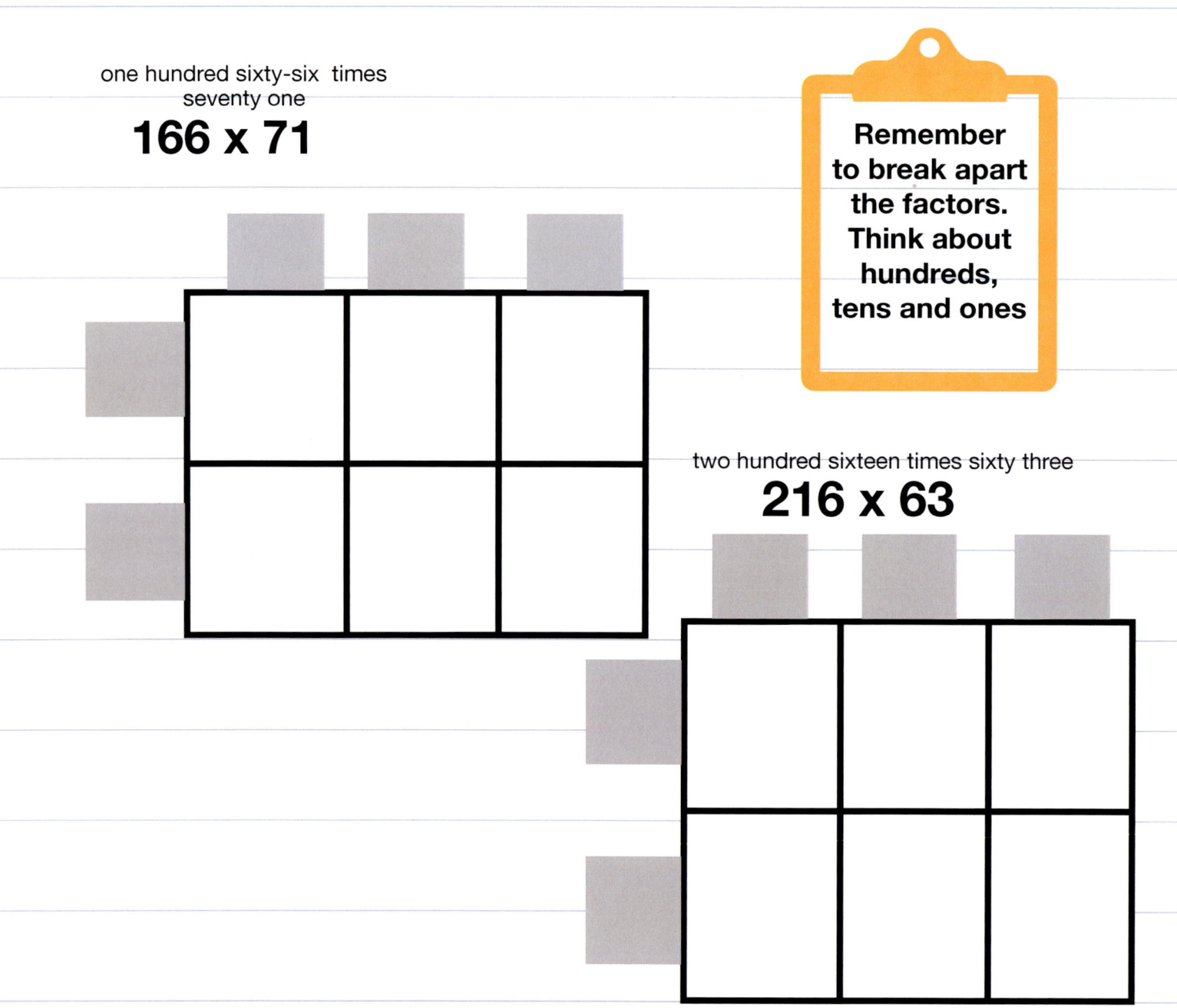

Multiply using Partial Products

fifty four times seventy three

ex.54 x 73

1: Multiply the ones.
3 x 4 = 12 then multiply
3 x 50 = 150.

2: Multiply the tens.
70 x 4 = 280 then
multiply 70 x 50 = 3500

3. Add all the
products together. (add
2 products at a time)

```
      54
  x  73
  ─────
      12
     150
     280
  + 3500
  ─────
    3942
```

final answer

```
     12
  +
    150
  ─────
    162

    280
  +
   3500
  ─────
   3780

   3780
  +
    162
  ─────
   3942
```

Multiply using Partial Products

four hundred twenty five times nine

ex. 425 x 9

1: Multiply the ones. 9 x 5 = 45

2: Multiply the tens. 9 x 20 = 180

3: Multiply 9 x 100 = 900

4. Add all the products together. (add 2 products at a time)

$$
\begin{array}{r}
425 \\
\times \quad 9 \\
\hline
45 \\
\overset{1}{180} \\
3600 \\
\hline
\mathbf{3825}
\end{array}
$$

$$
\begin{array}{r}
\overset{1}{45} \\
180 \\
\hline
225 \\
+3600 \\
\hline
3825
\end{array}
$$

final answer

Multiply using Lattice Multiplication

twenty three times sixty four

ex. 23 x 64

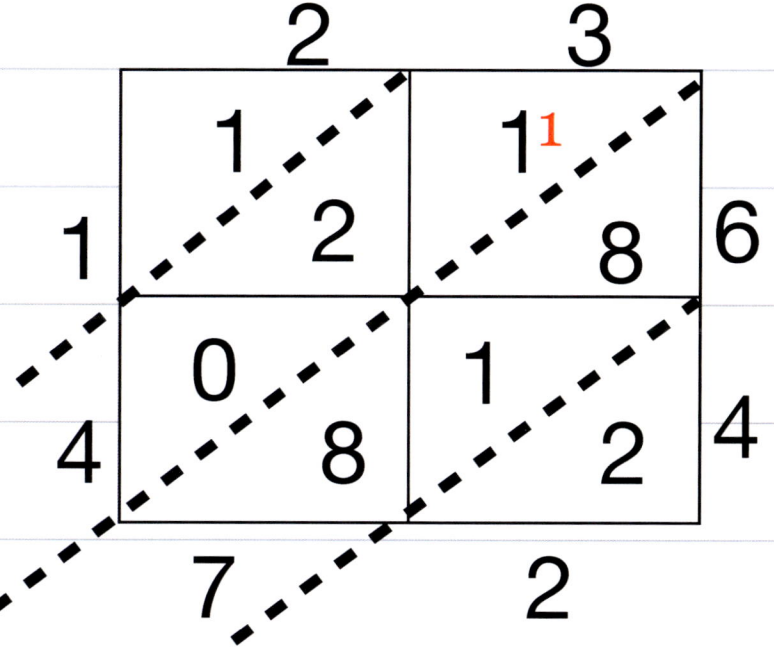

ex. 23 x 64 = 1472

1. Multiply 3 x 6 = 18. write the product in the top right box.

2. Multiply 3 x 4 = 12. write the product in the bottom right box.

3. Multiply 2 x 6 =12. write the product in the top left box.

4. Multiply 2 x 4 = 8. write the product in the bottom left box.

5. Add the numbers diagonally.

Multiply using Lattice Multiplication

two hundred three times forty three

ex. 223 x 43

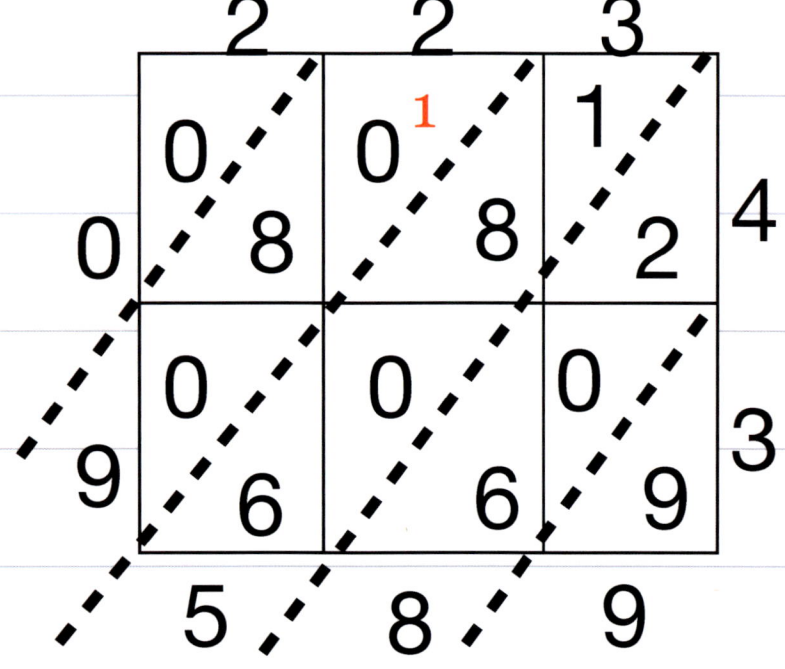

ex. 223 x 43 = 9589

1. Multiply 3 x 4 =12, 4 x 2 = 08, and 4 x 2 = 08. Put each product in the corresponding box

2. Multiply 3 x 3 = 09, 3 x 2 = 06, and finally 3 x 2 = 06.

3. Put each of the products in the corresponding box.

4. Find the sum of each diagonally row.

Let's Practice: Multiplying Whole Numbers using Lattice Multiplication

26 x 49

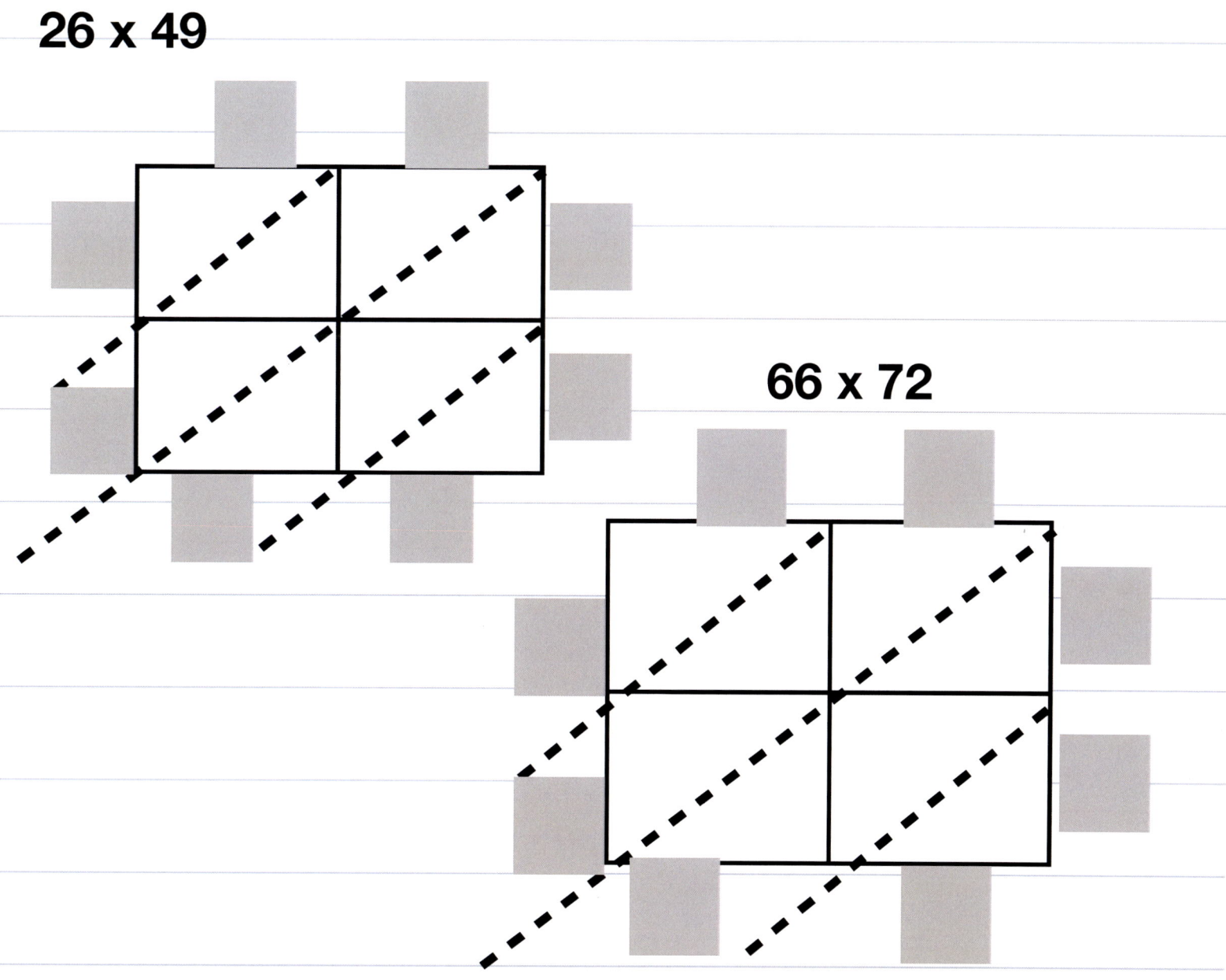

66 x 72

Let's Practice: Multiplying Whole Numbers using Lattice Multiplication

402 x 27

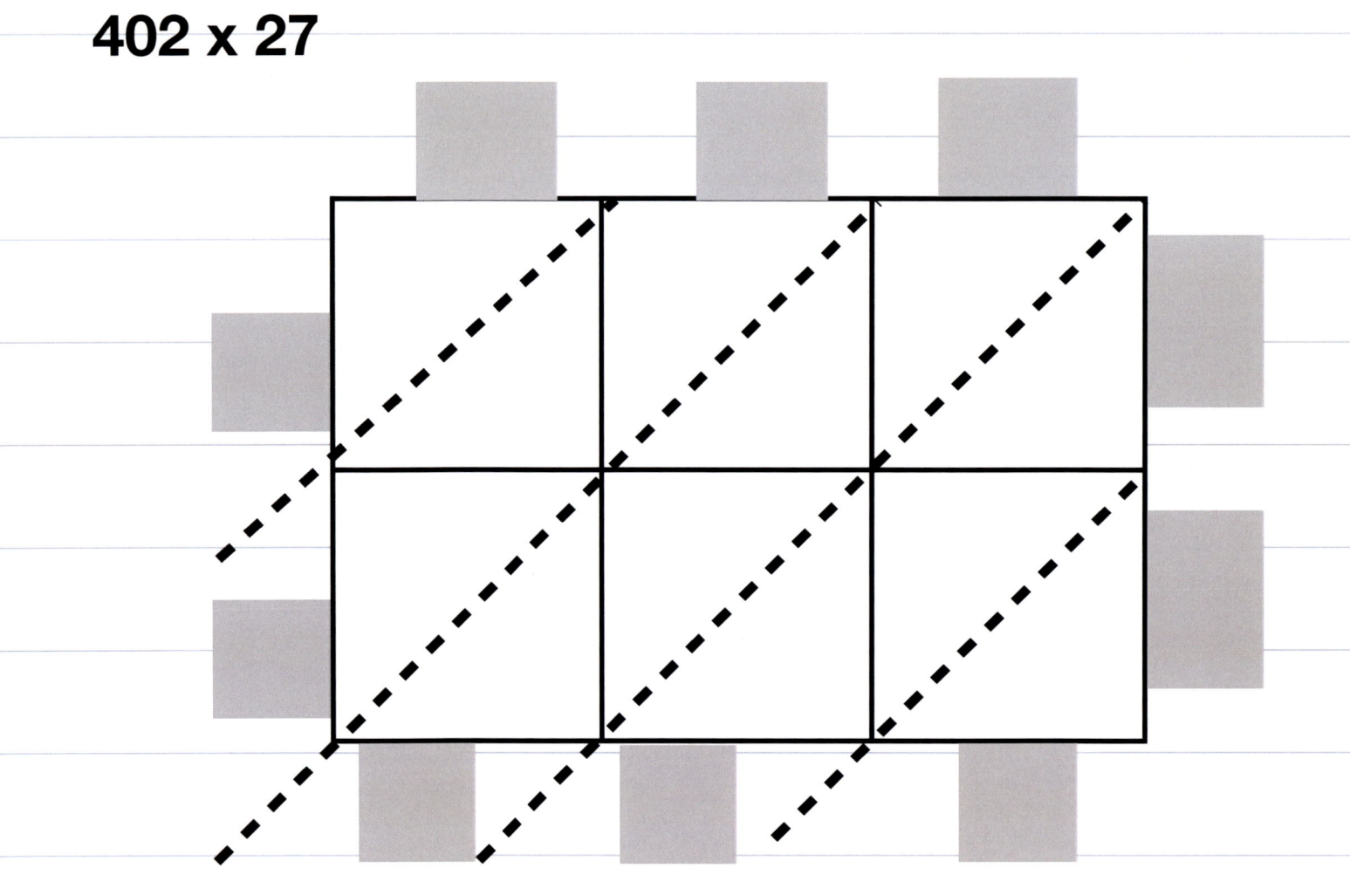

Multiply Whole Numbers using Standard Algorithm

forty six times twenty seven

ex. 46 x 27

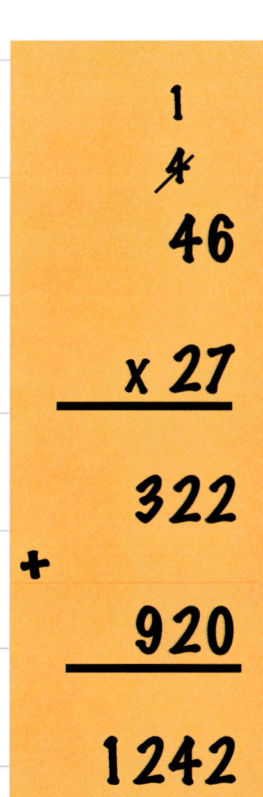

1. Start by multiplying the number in the ones place (multiplier) with all the top numbers (multiplicand) 7 x 6 = 42

2. Next multiply 7 x 4 = 28 + 4

3. Now we multiply the tens, put a 0 as the place holder in the ones place.

4. Now multiply the number in the tens place, with all the top numbers

5. Multiply 2 x 6 = 12

6. Multiply 2 x 4 = 8 + 1 = 9

7. Finally add the two amounts

46 x 27 = 1242

forty six times twenty seven equals
one thousand two hundred forty two

Multiply Whole Numbers using Standard Algorithm

three hundred forty six times twenty seven

ex. 346 x 27

$$\begin{array}{r} {\scriptstyle 1} \\ {\scriptstyle 3}\ \not{4} \\ 346 \\ \times\ 27 \\ \hline 2422 \\ {\scriptstyle +1} \\ 6920 \\ \hline 9342 \end{array}$$

1. Start by multiplying the number in the ones place (multiplier) with all the top numbers (multiplicand) 7 x 6 = 42

2. Next multiply 7 x 4 = 28 + 4

3. Then multiply 7 x 3 = 21 + 3

4. Now put a 0 as the place holder in the ones place.

5. Now multiply the number in the tens place, with all the top numbers

6. Multiply 2 x 6 = 12

7. Multiply 2 x 4 = 8 + 1 = 9

8. Multiply 2 x 3 = 6 + 3 = 9

9. Finally add the two amounts

346 x 27 = 9342

three hundred forty six times twenty seven equals nine thousand three hundred forty two

Multiplication of Whole Numbers

What I Learned

Multiplication of Decimals

Multiplication of Decimals

What I Know	What I want to Know

Decimal Place Value & Hundreds Chart

Decimals are smaller than whole numbers. Decimal numbers are written with a decimal point.

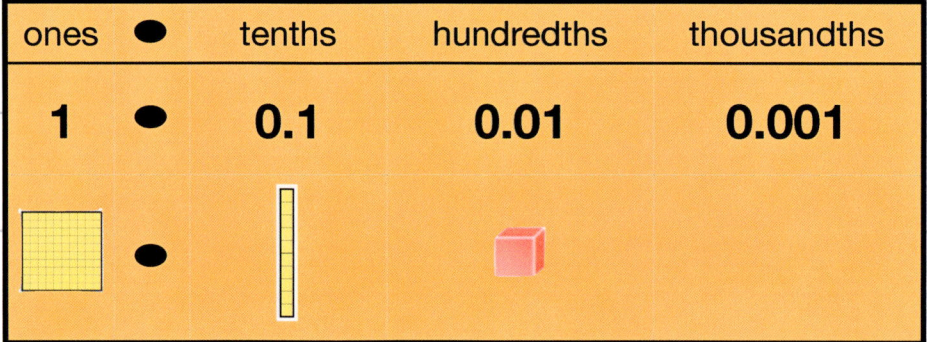

ones	•	tenths	hundredths	thousandths
1	•	0.1	0.01	0.001

0.01	0.02	0.03	0.04	0.05	0.06	0.07	0.08	0.09	0.10
0.11	0.12	0.13	0.14	0.15	0.16	0.17	0.18	0.19	0.20
0.21	0.22	0.23	0.24	0.25	0.26	0.27	0.28	0.29	0.30
0.31	0.32	0.33	0.34	0.35	0.36	0.37	0.38	0.39	0.40
0.41	0.42	0.43	0.44	0.45	0.46	0.47	0.48	0.49	0.50
0.51	0.52	0.53	0.54	0.55	0.56	0.57	0.58	0.59	0.60
0.61	0.62	0.63	0.64	0.65	0.66	0.67	0.68	0.69	0.70
0.71	0.72	0.73	0.74	0.75	0.76	0.77	0.78	0.79	0.80
0.81	0.82	0.83	0.84	0.85	0.86	0.87	0.88	0.89	0.90
0.91	0.92	0.93	0.94	0.95	0.96	0.97	0.98	0.99	1.00

Multiplying Decimals

A decimal is a fraction. Decimals are part of a whole. Decimals are written using a decimal point.

1 decimal place in the factor

$$\begin{array}{r} 2.1 \\ \times\ 7 \\ \hline 14.7 \end{array}$$

say, two and one tenths times seven

Notice there is one decimal place in the product

Did you know, we can multiply decimals the way we multiply whole numbers? Just count the amount of decimal places in the factor(s) Place the decimal in the same number of spaces in the product.

2 decimal places in the factor

Notice there are two decimal places in the product

$$\begin{array}{r} 2.13^{2} \\ \times\ \ 7 \\ \hline 14.91 \end{array}$$

say, two and thirteen hundredths times seven

3 decimal places in the factors

$$\begin{array}{r} {}^{2}\ \ {}^{1} \\ {}^{3}\ \ {}^{2} \\ 2.74 \\ \times\ \ \ 3.5 \\ \hline 1370 \\ +8220 \\ \hline 9.590 \end{array}$$

say, two and seventy-four hundredths times three and five tenths

Notice there are three decimal places in the product

Multiplying Decimals by Multiples of 10, 100, and 1000 using a place value chart

three and five tenths times ten

3.5 x 10

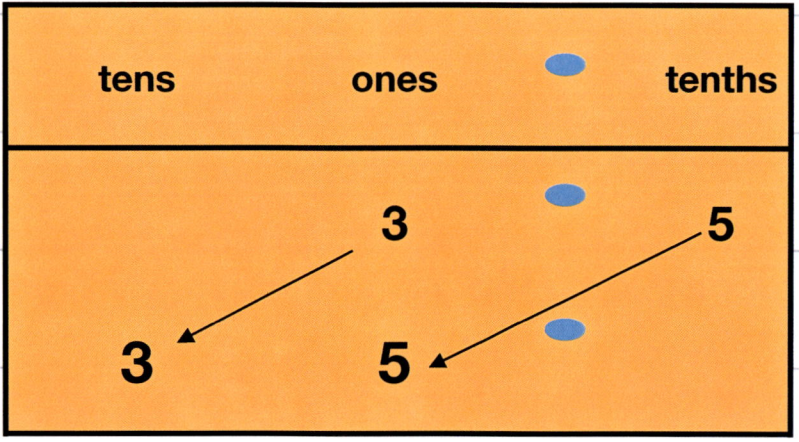

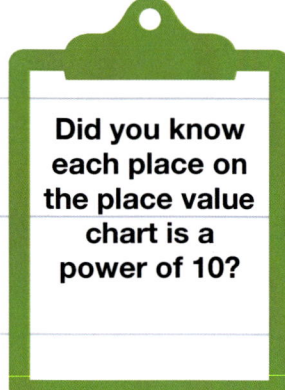

Did you know each place on the place value chart is a power of 10?

When we multiply by multiples of 10 we shift to the left ←
We shift left 1 place (times 10).
Start with the ones place. The 3 shifts from the ones place to the tens place. 3 times 10 is 30
The 5 shifts from the tenths place to the ones place. 5 tenths times 10 is 5.

3.5 x 10= 35

three and five tenths times ten
equals thirty five

Multiplying Decimals by Multiples of 10, 100, and 1000 using a place value chart

three and five tenths times one hundred

3.5 x 100

When we multiply by multiples of 10 we shift to the left ⟵
We shift left 2 place (times 100).
Start with the ones place. The 3 shifts from the ones place to the hundreds place. 3 times 100 is 300
The 5 shifts from the tenths place to the tens place. 5 tenths times 100 is 50.
The zero in the ones place is the place holder. The ones place should not be blank

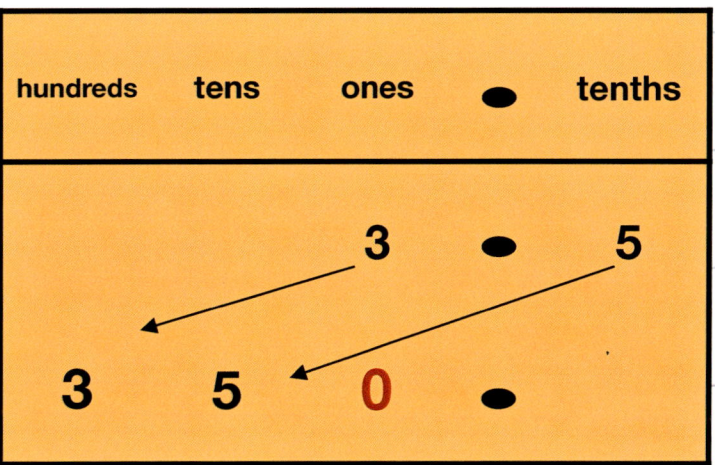

3.5 x 100 = 350

three and five tenths times one hundred equals three hundred fifty

Multiplying Decimals by Multiples of 10, 100, and 1000 using a place value chart

three and five tenths times one thousand

3.5 x 1000

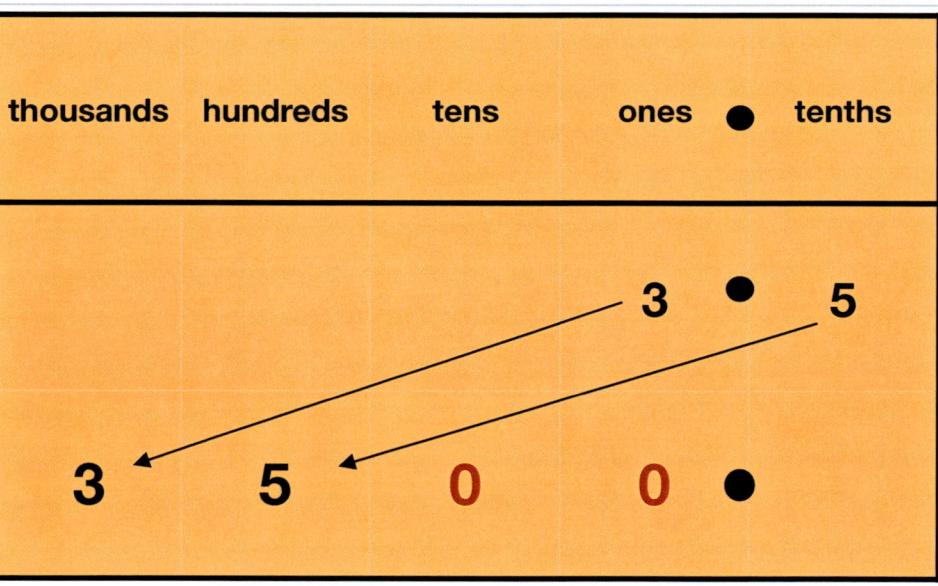

thousands	hundreds		tens	ones •	tenths
				3 •	5
3	5		0	0 •	

When we multiply by multiples of 10 we shift to the left ←
We shift left 3 place (x 1000).
Start with the ones place. The 3 shifts from the ones place to the thousands place. 3 times 1000 is 3000
The 5 shifts from the tenths place to the hundreds place. 5 tenths times 100 is 500.
The zero in the ones place is the place holder. The ones place should not be blank
The zero is a place holder in the tens place also.

3.5 x 1000 = 3500

three and five tenths times one thousand equals three thousand five hundred

Multiplying Decimals using Visuals & Place Value Chart

two and three tenths times three

ex. 2.3 x 3

Here is a place value chart 3 groups of two thirds

Ones	•	Tenths
2	•	3
2	•	3
2	•	3

Here is a visual of 3 groups of two thirds

You can count the pictures to find your final answer

6	•	9
six	and	nine tenths

2.3 x 3 = 6.9

two and three tenths times three
equals six and nine tenths

Multiplying Decimals using Visuals & Place Value Chart

forty two and thirty six hundredths times three

ex. 42.36 x 3

Hundr eds	Tens	Ones	●	Tenths	Hundre dths
		1		1	
	4	2	●	3	6
	4	2	●	3	6
	4	2	●	3	6
1	2	7	●	0	8

Hundr eds	Tens	Ones	●	Tenths	Hundreths
	regroup 1 ten to 1 hundred		and	regroup 1 tenth to 1 one	regroup 10 hundredths to 1 tenth
1	2	7	and	0	8

Did you know to regroup means to rename the value. Every set of ten, we regroup

42.36 x 3 = 127.08

forty two and thirty six hundredths times three equals one hundred twenty seven and eight tenths

Multiplying Decimals using Tape Diagrams & Grid Models

eight tenths tenths times nine

ex .8 x 9

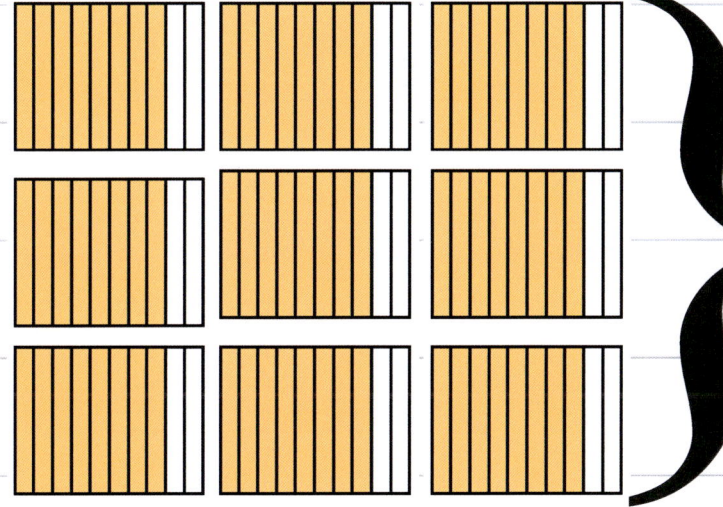

Visual Model - nine sets of eight tenths

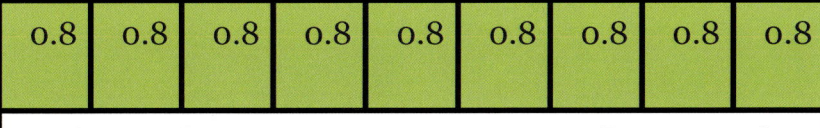

Tape Diagram

eight tenths times nine = seven and two tenths

.8 x 9 = 7.2

eight tenths tenths times nine equals
seven and two tenths

Multiplying Decimals using Tape Diagrams & Grid Models

ex **.7 x 16**

seven tenths times sixteen

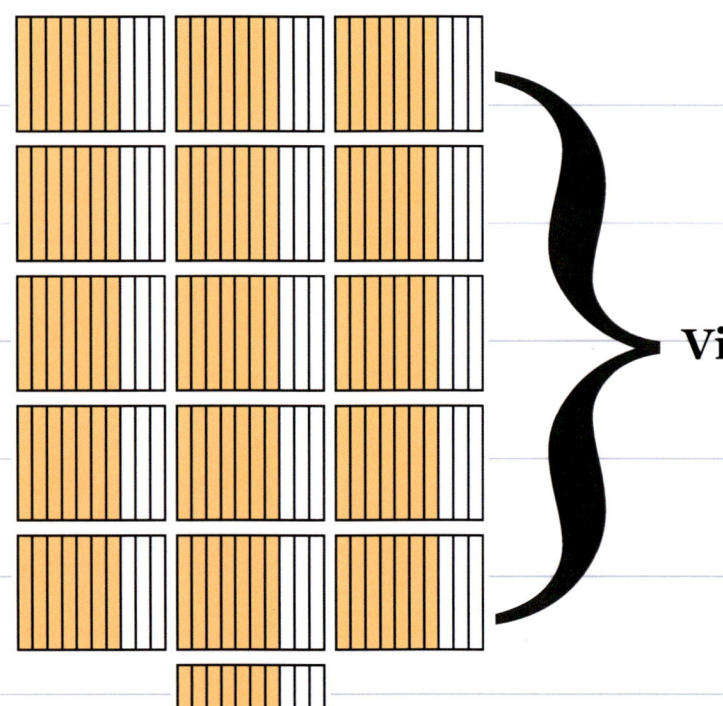

Visual Model

.7 x 16 = 11.2

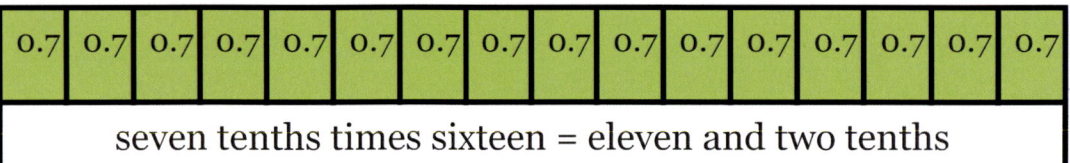

seven tenths times sixteen = eleven and two tenths

Tape Diagram

Multiplying Decimals using Box Models

three and fifteen hundredths times twenty four

ex. 3.15 x 24

Multiply each column with each row.
Add the products for the final answer.

	3	0.1	0.05
20	20 x 3 = 60	20 x .1 = .20	20 x 0.005 = 1000
4	12	4 x .1= 0.4	4 x .05= .20

There are two decimal places in the factor, there should be two decimal places in the product

$$
\begin{array}{r}
63.00 \\
+12.60 \\
\hline
75.60
\end{array}
$$

3.15 x 24 = **75.60**

$$
\begin{array}{r}
^1_2 \\
3.15 \\
\times\ 24 \\
\hline
1260 \\
+\ 6300 \\
\hline
7560
\end{array}
$$

final answer

Multiplying Decimals using Box Models

four and seven tenths times eight

ex 4.7 x 8

Think about multiplying whole numbers. Remove the decimals. Multiply each column with each row.
Add the products for the final answer.

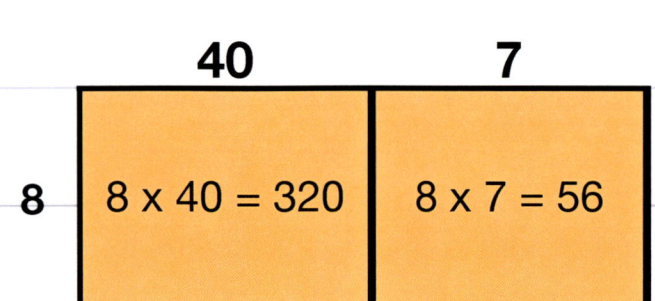

	40	7
8	8 x 40 = 320	8 x 7 = 56

```
   320
+   56
   376
```

There are one decimal place in one of the factors, there should be one decimal places in the product.

4.7 x 8 = **37.6**

Multiplication of Decimals

What I Learned

Multiplication of Fractions

Multiplication of Fractions

What I Know	What I want to Know

Multiplying Fractions by Whole Numbers using Counters

When multiplying fractions by whole numbers, you should convert the whole number into a fraction by using 1 as the denominator. Multiply the numerators then multiply the denominators.

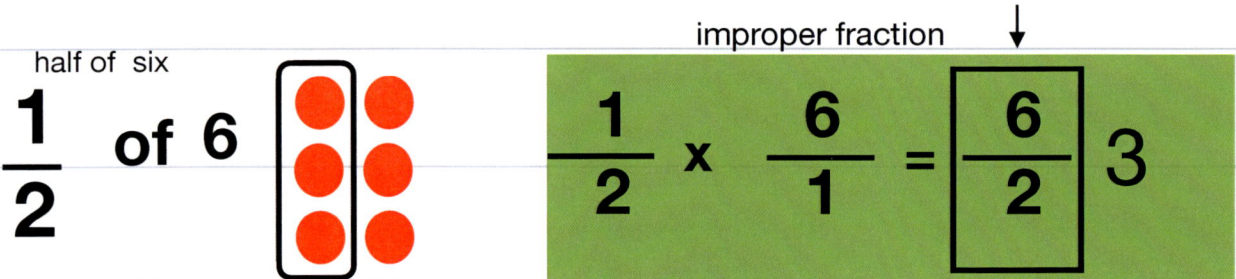

half of six

$$\frac{1}{2} \text{ of } 6$$

improper fraction ↓

$$\frac{1}{2} \times \frac{6}{1} = \frac{6}{2} \quad 3$$

half times six

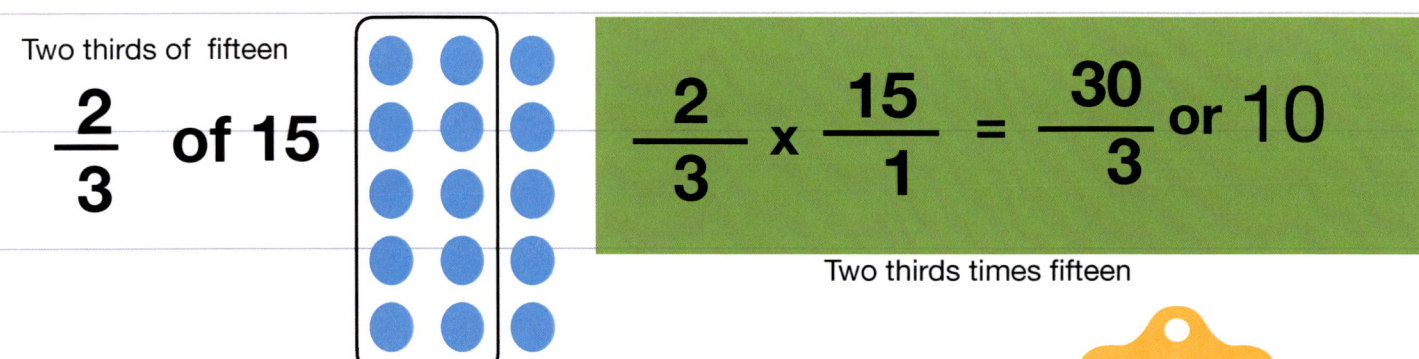

Two thirds of fifteen

$$\frac{2}{3} \text{ of } 15$$

$$\frac{2}{3} \times \frac{15}{1} = \frac{30}{3} \text{ or } 10$$

Two thirds times fifteen

Think about how many rows in total Think about how many counter are in each row Look at the counters that are circled

Multiplying Fractions by Whole Numbers using Counters

three fourths of twenty eight

$$\frac{3}{4} \text{ of } 28$$

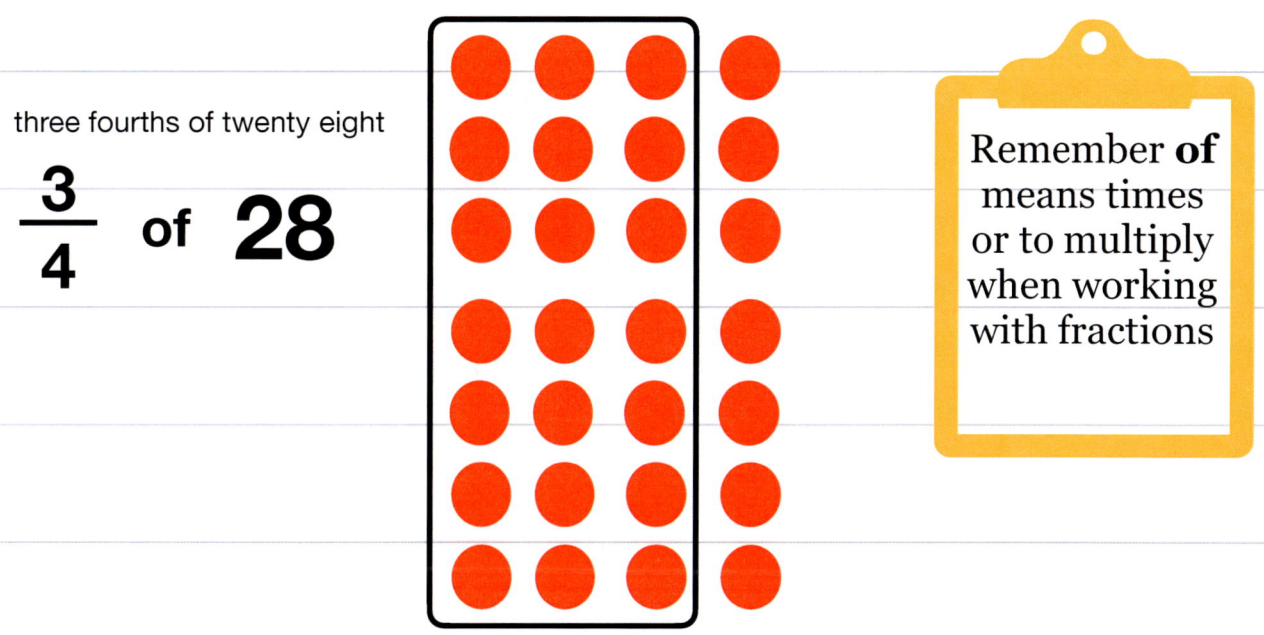

Remember **of** means times or to multiply when working with fractions

three fourths times twenty eight

$$\frac{3}{4} \times \frac{28}{1} = \frac{84}{4} \text{ or } 21$$

improper fraction

To convert an improper fraction to a whole number, we divide 84 ÷ 4. We can make 21 groups of 4 with 84

Multiplying Fractions by Whole Numbers Multiple Representations

Multiply fraction by Whole Numbers

1. Make the whole number a fraction by using 1 as the denominator.

2. Multiply the numerator then Multiply the denominator

3. Improper fraction, convert to mixed number - reduce if necessary.

$$\frac{1}{2} \times 3 \quad \boxed{\text{or}} \quad \frac{1}{2} + \frac{1}{2} + \frac{1}{2}$$

Repeated Addition   one half times three

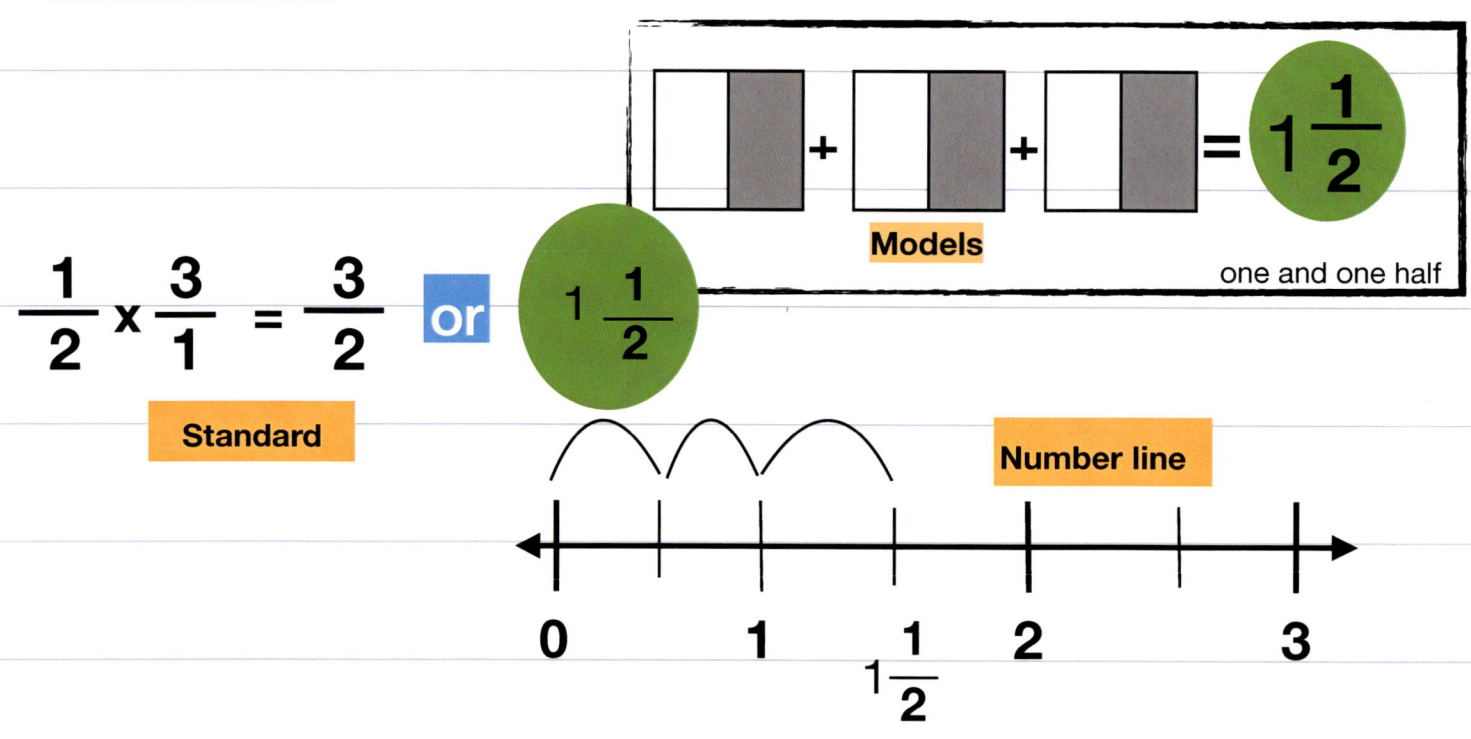

Models

one and one half

$$\frac{1}{2} \times \frac{3}{1} = \frac{3}{2} \quad \boxed{\text{or}} \quad 1\frac{1}{2}$$

Standard

Number line

Multiplying Fractions by Whole Numbers Multiple Representations

1. Make the whole number a fraction by using 1 as the denominator.
2. Multiply the numerator then Multiply the denominator
3. Improper fraction, convert to mixed number - reduce if necessary.

Two thirds times four

$$\frac{2}{3} \times 4 \quad \text{or} \quad \frac{2}{3} + \frac{2}{3} + \frac{2}{3} + \frac{2}{3}$$

Repeated Addition

$$= 2\frac{2}{3} \quad \text{or}$$

models

$$\frac{2}{3} \times \frac{4}{1} = \frac{8}{3} \quad \text{or} \quad 2\frac{2}{3}$$

Standard

Each jump is two thirds of a jump

0   1   2 $2\frac{2}{3}$ 3   4

number line

Multiplying Fractions by Fractions

Unit Fraction by Unit Fraction

$$\frac{1}{2} \times \frac{1}{3}$$

half times one third

👣 1. Start by multiplying the numerators

👣 2. Multiply the denominators

half model

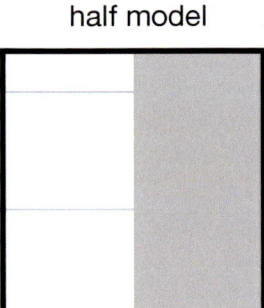

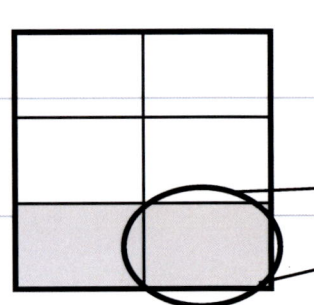

Then we cut the square into 3 equal size parts horizontally

This is the part that is double shaded or where the parts overlap.
There are 6 total parts.
One of the parts are double shaded.

$$\frac{1}{2} \times \frac{1}{3} = \frac{1}{6}$$

This works because we draw a square with 2 vertical columns.

Multiplying Fractions by Fractions

Unit Fraction by Non Unit Fraction

$$\frac{1}{2} \times \frac{2}{3}$$

half times two thirds

👣 1. Start by multiplying the numerators

👣 2. Multiply the denominators

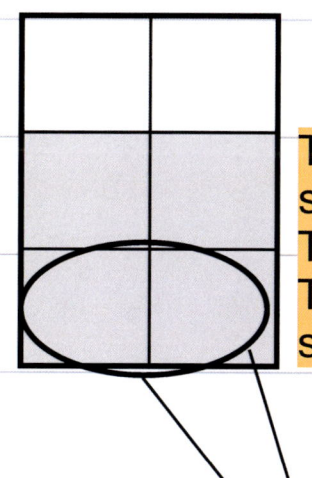

This is the part that is double shaded.
There are 6 total parts.
Two of the parts are double shaded.

1. This works because we draw a square with 2 vertical columns.

2. Then we cut the square into 3 equal size parts horizontally

$$\frac{1}{2} \times \frac{2}{3} = \frac{2}{6} = \frac{1}{3} \text{ reduce}$$

Multiplying Fractions by Fractions

Non Unit Fraction by Non Unit Fraction

$$\frac{2}{3} \times \frac{3}{4}$$

two thirds times three fourths

👣 1. Start by multiplying the numerators

👣 2. Multiply the denominators

👣 3. Simply if needed

1. This works because we draw a square with 2 vertical columns.

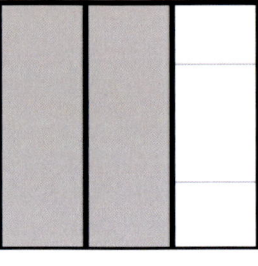

2. Then we cut the square into 4 equal size parts horizontally

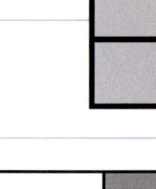

This is the part that is double shaded.
There are 12 total parts.
six of the parts are double shaded.

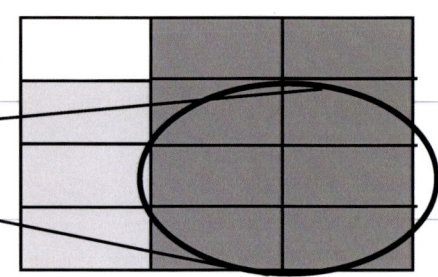

$$\frac{2}{3} \times \frac{3}{4} = \frac{6}{12} \text{ or } \frac{1}{2} \text{ reduce}$$

Multiplying Mixed Numbers by Whole Numbers

Convert the mixed number to an improper fraction, then make the whole number a fraction. Use 1 as the denominator

$2\dfrac{1}{2}$ x 4

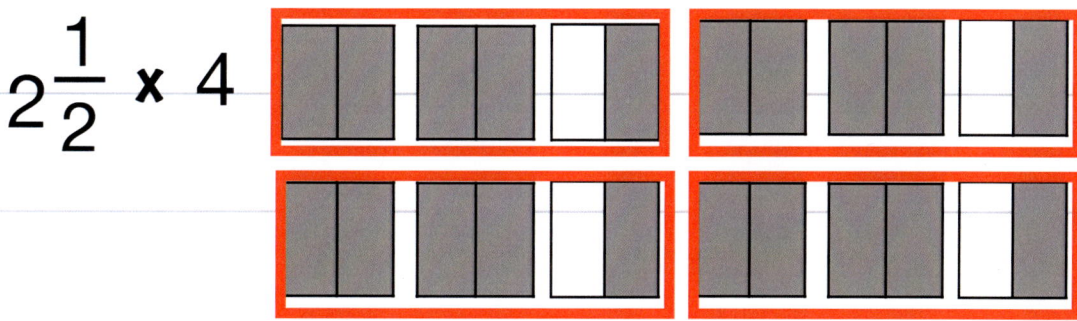

When you convert a mixed number to an improper fractions, multiply the denominator with the whole number, then add the numerator

$2\dfrac{1}{2}$ x 4 $=$ $\dfrac{5}{2}$ x $\dfrac{4}{1}$ $=\dfrac{20}{2}$ $=$ 10

two and one half converted to an improper fraction is five halves

Convert the improper fraction back to a mixed number or whole number after you multiply

Multiplication of Fractions

What I Learned

Division of Whole Numbers

Division of Whole Numbers

What I Know	What I want to Know

Division Concepts/Basics

To divide means to separate, share out equally into groups.
Division is the inverse or opposite of multiplication.

Ways we see division: $12 \div 4$ or $4\overline{)12}$ or $\dfrac{12}{4}$

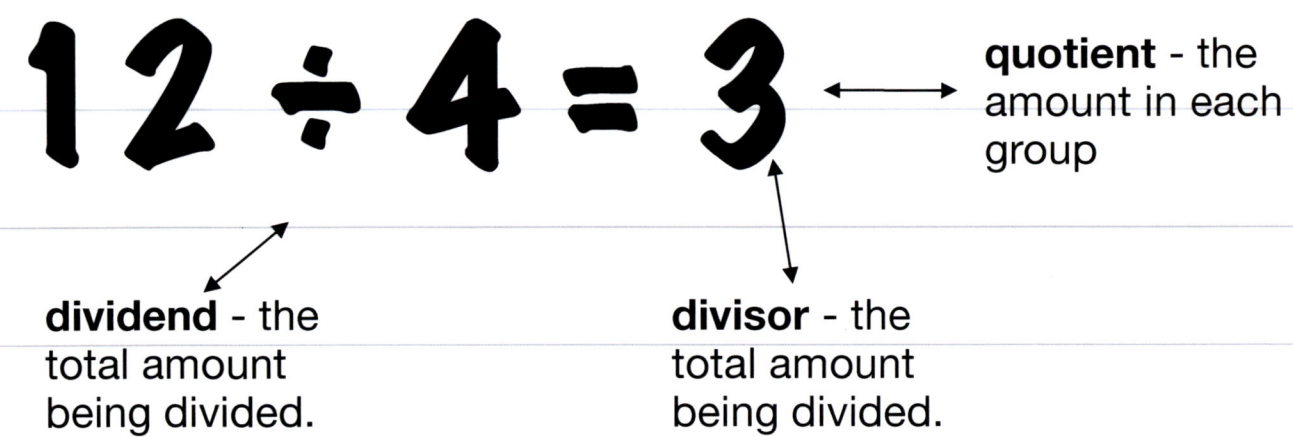

$$12 \div 4 = 3$$

quotient - the amount in each group

dividend - the total amount being divided.

divisor - the total amount being divided.

Think about sharing 12 computers equally among four classrooms. Each classroom will receive 3 computers

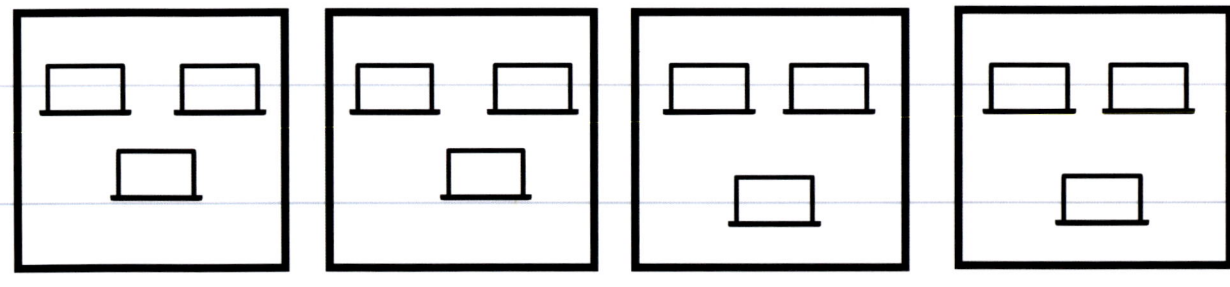

Division using a Multiplication Chart

fifty six divided by eight

ex. 56 ÷ 8

👣 1. Find the dividend on the multiplication chart

👣 2. Move up along the blue shaded area to find the divisor

👣 3. Move left to find the quotient

1	2	3	4	5	6	7	8	9	10	11	12
2	4	6	8	10	12	14	16	18	20	22	24
3	6	9	12	15	18	21	24	27	30	33	36
4	8	12	16	20	24	28	32	36	40	44	48
5	10	15	20	25	30	35	40	45	50	55	60
6	12	18	24	30	36	42	48	54	60	66	72
7	14	21	28	35	42	49	56	63	70	77	84
8	16	24	32	40	48	56	64	72	80	88	96
9	18	27	36	45	54	63	72	81	90	99	108
10	20	30	40	50	60	70	80	90	100	110	120
11	22	33	44	55	66	77	88	99	110	121	132
12	24	36	48	60	72	84	96	108	120	132	144

56 ÷ 8 = 7

fifty six divided by eight equal seven

Dividing Whole Number using Equal Groups

twelve divided by four

12 ÷ 4

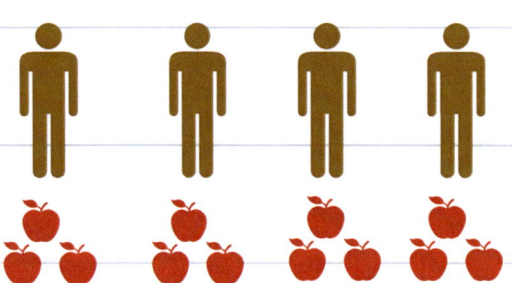

Twelve apples shared among 4 people. each person get 3 apples

12 ÷ 4 = 3

thirty six divided by 4

36 ÷ 4

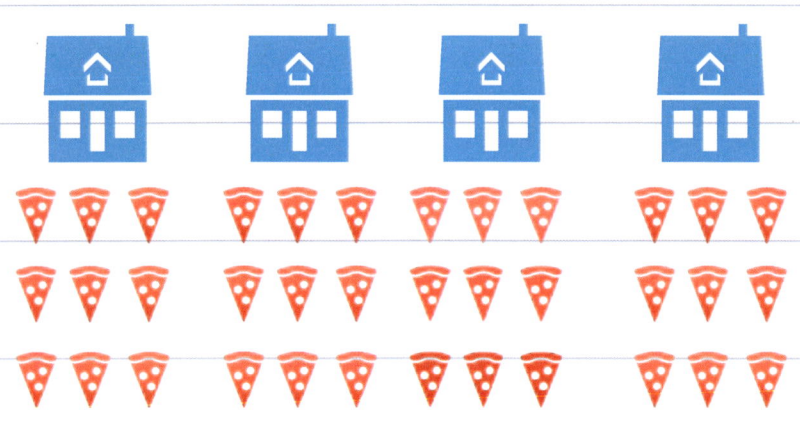

Thirty six box of pizzas shared among 4 classes. Each class get 9 boxes of pizza.

36 ÷ 4 = 9

Dividing Whole Number using Equal Groups

forty eight divided by eight

48 ÷ 8

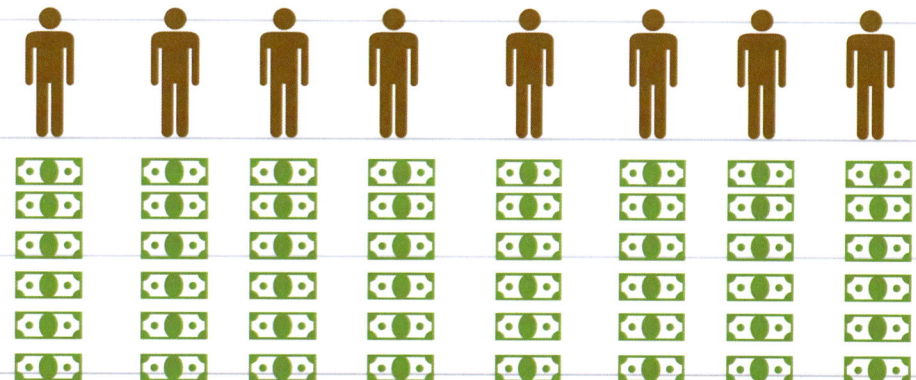

thirty six dollars shared amount eight people

48 ÷ 8 = 6

forty eight divided by eight equals six

Dividing 2 digits by 1 digit
Whole Numbers
no remainders
using multiplication

forty eight divided by four

$$48 \div 4$$

I know $4 \times 2 = 8$.

I also know $4 \times 10 = 40$

so,

$8 + 40 = 48$.

that means $48 \div 4 = 12$

Use friendly numbers like 1, 2, 5, 10, and even 20

Multiplying Up Strategy

$4 \times 5 = 20$ ⟩ 40

$4 \times 5 = 20$ ⟩ 48

$4 \times 2 = 8$

$5 + 5 + 2 = 12$

$4 \times 10 = 40$ ⟩ 44

$4 \times 1 = 4$ ⟩ 48

$4 \times 1 = 4$

$10 + 1 + 1 = 12$

$$48 \div 4 = 12$$

forty eight divided by four equals twelve

Dividing 2 digits by 1 digit Whole Numbers

with remainders using multiplication

twenty two divided by three

ex. 22 ÷ 3

Use friendly numbers like 1, 2, 5, 10, and even 20

$3 \times 2 = 6$

$3 \times 5 = 15$

21

There is 1 left over. We can't make another group of 3.

$3 \times 2 = 6$

$3 \times 2 = 6$

12

$3 \times 2 = 6$

18

$3 \times 1 = 3$

21

Add the highlighted numbers to find the quotient

I know $3 \times 5 = 15$.

I also know $3 \times 2 = 6$

$15 + 6 = 21$ since I can't make another group of 3 then I know there are 7 groups of 3 in 22

*Add the multipliers to find the quotient

$22 \div 3 = 7 \text{ R}1$

twenty two divided by three equals seven remainder one

Dividing 2 digits by 1 digit Whole Numbers

no remainders using multiplication

seventy eight divided by three

ex. 78 ÷ 3

$3 \times 10 = 30$

$3 \times 10 = 30$

$3 \times 6 = 18$

Think $3 \times$ ___ $= 78$. So let's multiply until the products add up to **78**.

Add the factors $10 + 10 + 6$ to find the quotient

$10 + 10 + 6 = 26$

$78 \div 3 = 26$

seventy eight divided by three
equals twenty six

Dividing 2 digits by 1 digit Whole Numbers

no remainders using quick pics

seventy eight divided by three

ex. 78 ÷ 3

Let **|** represent tens.
Let ● represent ones.

Make 3 groups. distribute the tens equally among the three groups

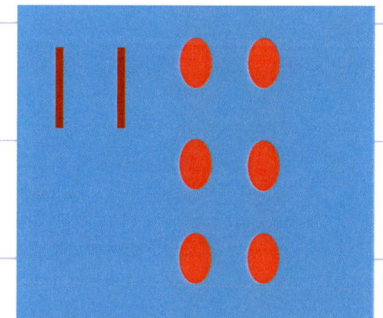 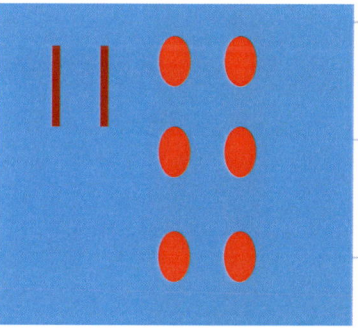 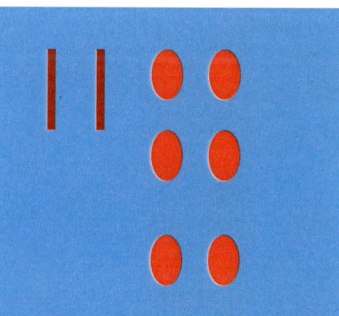

There are two **|**
and six ●
in each group.

Change the remaining ten into 10 ones. continue distributing with the ones.

| = ●●●●● ●●●●●

$$78 \div 3 = 26$$

seventy eight divided by three equals twenty six

Dividing 2 digits by 1 digit
Whole Numbers using repeated subtraction

seventy eight divided by three

ex. 78 ÷ 3

> **Write out the multiples of 3 using friendly numbers like 1, 2, 5, 10 or 20**

one group of 3 ->1 x 3 = 3

two groups of 3 ->2 x 3 = 6

five groups of 3 -> 5 x 3 = 15

ten groups of 3 -> 10 x 3 = 30

twenty groups of 3 -> 20 x 3 = 60

Subtract the largest groups of 3. Add up the subtracted groups for the quotient.

20	5	1
78	18	3
-60	-15	-3
18	3	0
20 + 5 + 1 = 26		

78 ÷ 3 = 26

seventy eight divided by three equals twenty six

Dividing 2 digits by 1 digit Whole Numbers

with remainders using quick pics

fifty four divided by four

ex. 54 ÷ 4

Let **|** represent tens.

Let ● represent ones.

Make 4 groups. distribute the tens equally among the four groups. then the ones

| Change this remaining ten into 10 ones. continue distributing with the ones.

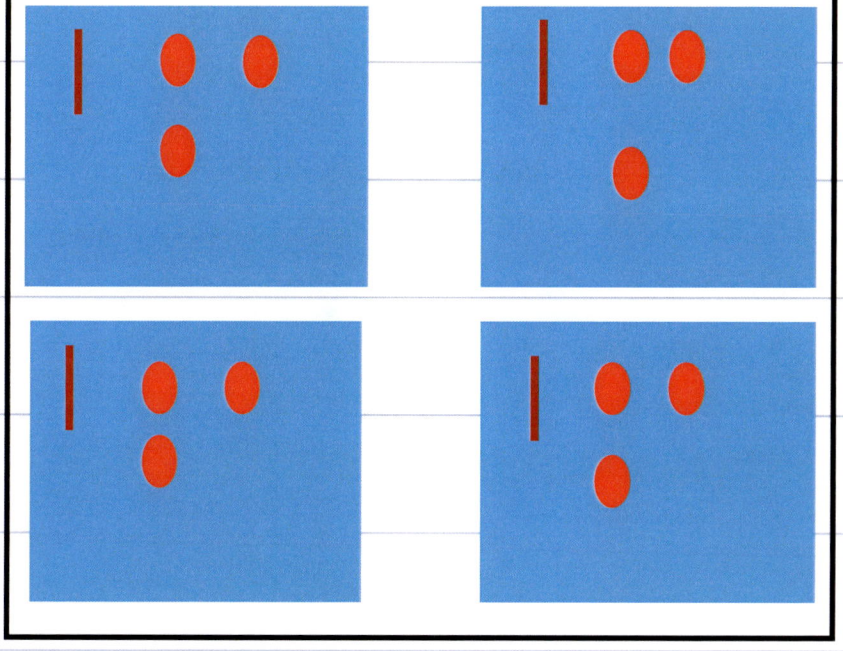

There is one **|**

and three ●

in each group.

There are two ● remaining

54 ÷ 4 = 13 r 2

fifty four divided by four equals thirteen remainder two

Dividing 2 digits by 1 digit
Whole Numbers

**using repeated subtraction
no remainders**

fifty four divided by four

ex. 54 ÷ 4

Write out the multiples of 4 using friendly numbers like 1, 2, 5, 10

one group of 4 ->1 x 4 = 4

two groups of 4 ->2 x 4 = 8

five groups of 4 -> 5 x 4 = 20

ten groups of 4 -> 10 x 4 = 40

10	2	1
54	14	6
-40	-8	-4
14	6	2
10 + 2 + 1 = 13		

Subtract the largest groups of 4. Add up the subtracted groups for the quotient.

54 ÷ 4 = 13 r 2

fifty four divided by four equals thirteen
remainder two

Dividing 2 digits by 1 digit Whole Numbers using Partial Quotients

eighty four divided by six

ex. 84 ÷ 6

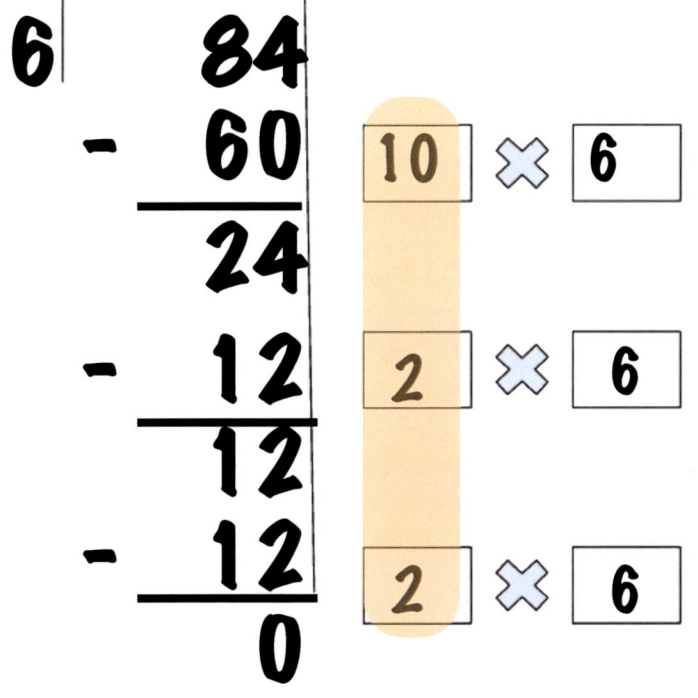

Use friendly numbers to multiply the divisor by.

$$84 \div 6 = \boxed{14}$$

eighty four divided by six equals fourteen

$$10 + 2 + 2 = 14$$

Dividing 2 digits by 2 digit Whole Numbers using Proportional Reasoning

eighty four divided by twelve

ex. 84 ÷ 12

Use a multiple of the dividend and divisor. Divide to the smallest multiple

$$84 \div 2 = 42$$

$$41 \div 2 = 21$$

$$12 \div 2 = 6$$

$$6 \div 2 = 3$$

$$21 \div 3 = 7$$

When you have reached the point you can not divide by the multiple, divide those factors with each other.

$$84 \div 12 = 7$$

Dividing 3 digits by 1 digit Whole Numbers

two hundred fifty eight divided by six

ex. 258 ÷ 6

Write out the multiples of 6 using friendly numbers like 1, 2, 5, 10

one group of 6 ->1 x 6 = **6**

two groups of 6 ->2 x 6 = **12**

five groups of 6 -> 5 x 6 = **30**

ten groups of 6 -> 10 x 6 = **60**

twenty groups of 6 -> 20 x 6 = **120**

Subtract the largest groups of 6. Add up the subtracted groups for the quotient.

20	20	2	1
258	138	18	6
-120	-120	-12	-6
138	18	6	0

20 + 20 + 2 + 1 = 43

258 ÷ 6 = 43

two hundred fifty eight divided by six equals forty three equals forty three

Dividing 3 digits by 1 digit Whole Numbers

two hundred fifty eight divided by six
ex. 258 ÷ 6

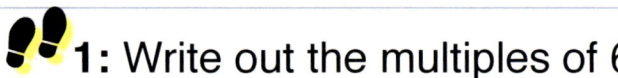 **1:** Write out the multiples of 6

one group of 6 ->1 x 6 = 6

two groups of 6 ->2 x 6 = 12

three groups of 6 -> 3 x 6 = 18

four groups of 6 -> 4 x 6 = 2

2:

Ask yourself: <u>Can 6 go into 2?</u> **No**

<u>Can 6 go into 25?</u> **Yes.** How many times? based on the multiples.

<mark>Divide ~ Multiply ~Subtract ~ Repeat</mark>

Divide - 6 can go into 25? about 4 times

Multiply - 6 x 4 = 24

Subtract - 25 - 24 = 1

Bring down the 8

and Repeat

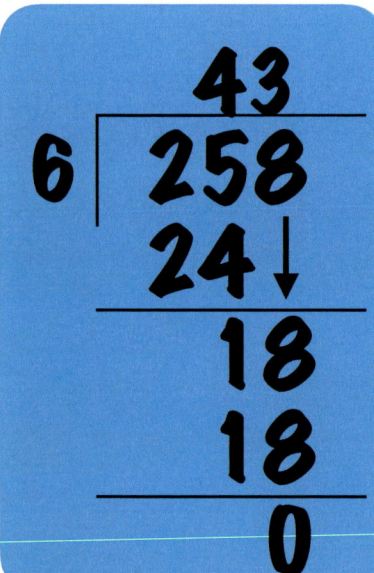

258 ÷ 6 = 43

Dividing 3 digits by 1 digit Whole Numbers

two hundred fifty eight divided by six

ex. 258 ÷ 6

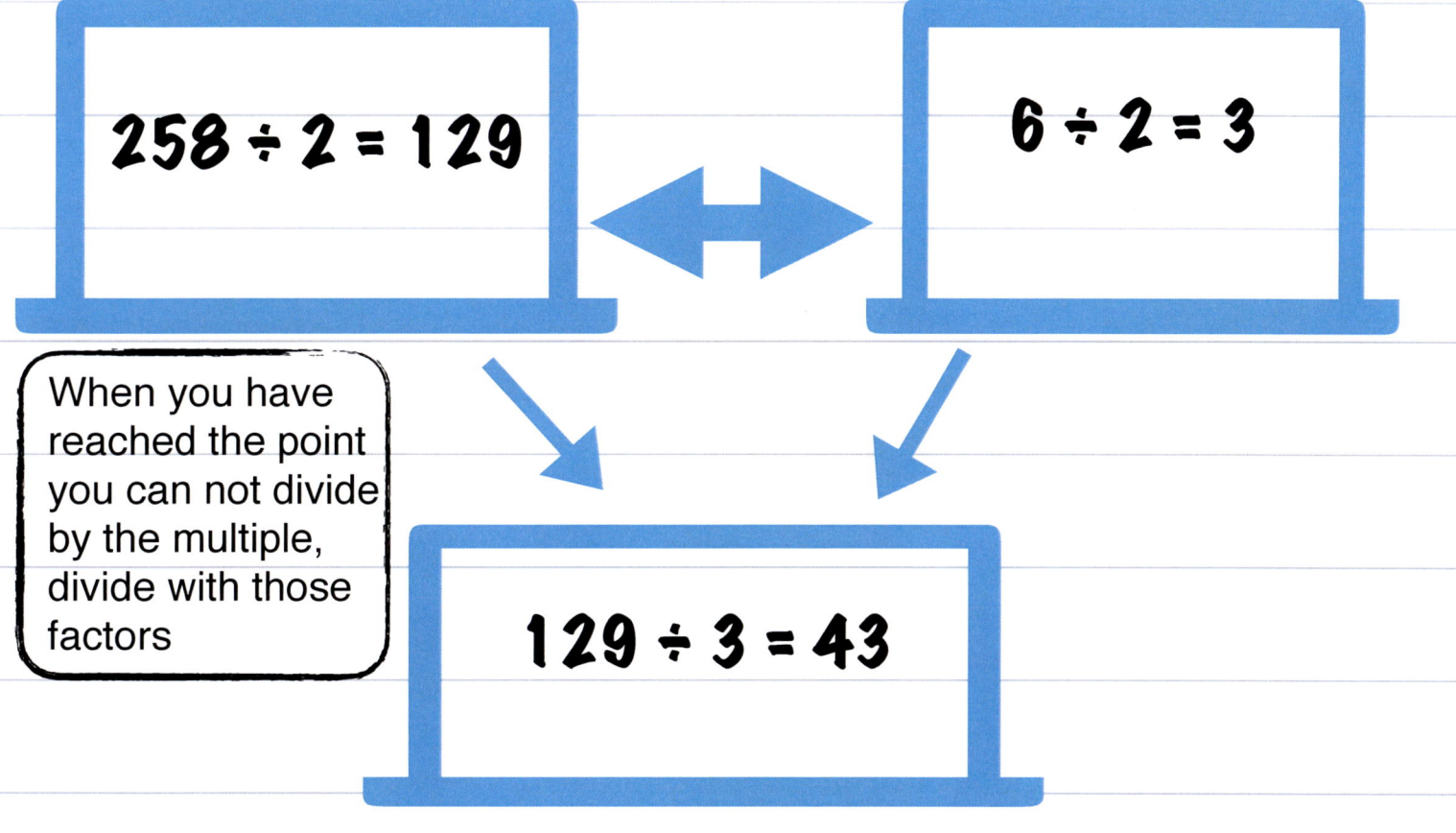

$$258 \div 2 = 129$$

$$6 \div 2 = 3$$

When you have reached the point you can not divide by the multiple, divide with those factors

$$129 \div 3 = 43$$

ex. 258 ÷ 6

two hundred fifty eight divided by six
equals forty three

Division of Whole Numbers

What I Learned

Division of Decimals

Division of Decimals

What I Know	What I want to Know

Dividing by Multiples of 10, 100, & 1000

3.5 ÷ 10

three and five tenths divided
by ten

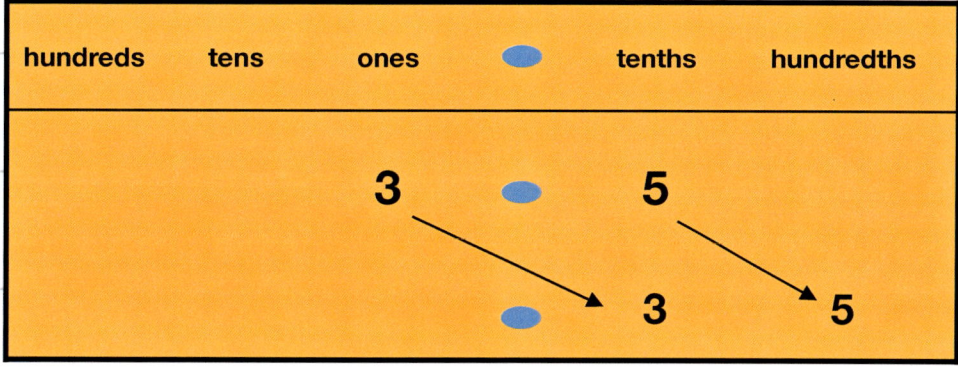

When we divide by multiples of 10 we shift to the right ⟶
We shift right 1 place (times 10).
Start with the tenths place. The 5 shifts from the tenths place
to the hundredths place. 5 tenths divided by 10 is five
hundredths
The 3 shifts from the ones place to the tenths place. 3 divided
by 10 is 3 tenths.

3.5 ÷ 10= .35

three and five tenths divided by
ten equal thirty five hundredths

Dividing by Multiples of 10, 100, & 1000

3.5 ÷ 100

three and five tenths divided by one hundred

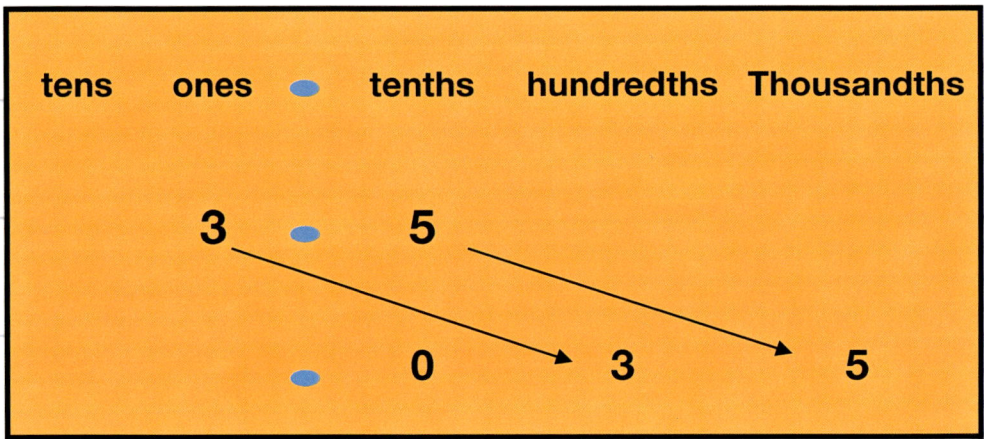

When we divide by 100 we shift to the right ——→
We shift right 2 place (times 100).
Start with the tenths place. The 5 shifts from the tenths place to the thousandths place. 5 tenths divided by 100 is five tenth
The 3 shifts from the ones place to the tenths place. 3 divided by 10 is 3 tenths. Put a zero in the tenths place.

3.5 ÷ 100 = .035

three and five tenths divided by one hundred equals thirty five thousandths

Dividing by Multiples of 10, 100, & 1000

3 ÷ 1000

there divided by one thousand

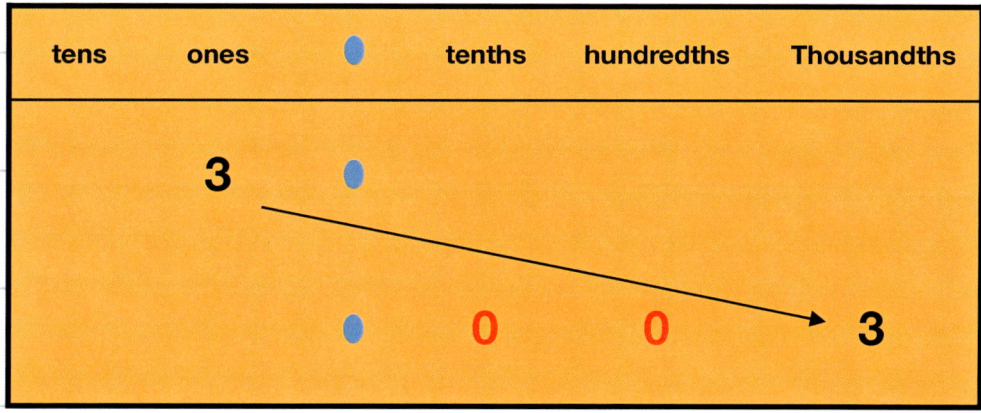

When we divide by 1000 we shift to the right ⟶
We shift right 3 places (times 1000).
Start with the ones place. The 3 shifts from the ones place to
the thousandths place.
The tenths and hundredths place should not be blank.
The zero is a place holder for the tenths and hundredths
place.

3 ÷ 1000 = .003

three divided by one thousand equals three
thousandths

Dividing Decimals

Dividing decimals are easy, by making them into whole numbers

Remember

$$\text{Divisor} \overline{\smash{)}\ \overset{\text{Quotient}}{\text{Dividend}}}$$

Dividend ÷ Divisor = Quotient

Dividing Decimals using Place Value

4 groups of _____ tenths is 1.6

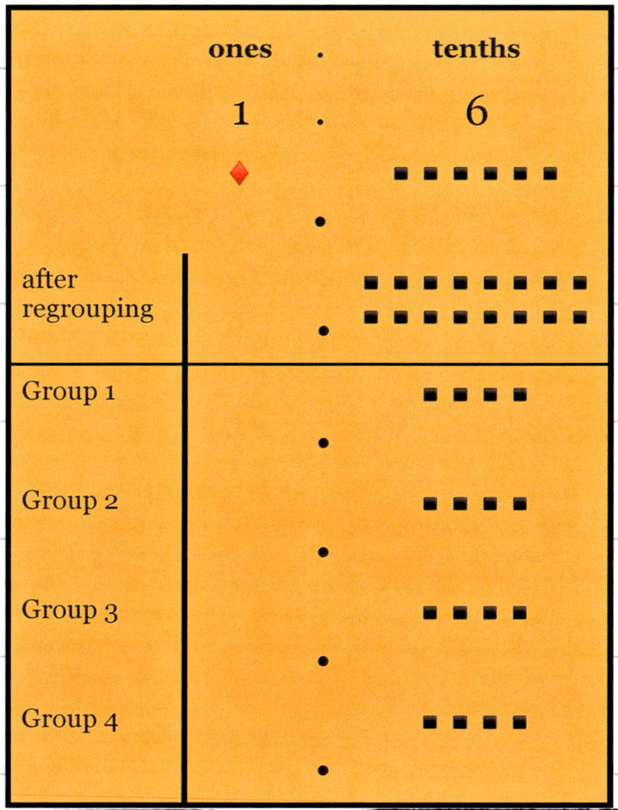

ones	.	tenths
1	.	6

after regrouping

Group 1

Group 2

Group 3

Group 4

Remember
1 tenths = 10 tenths
so 1.6 = 16 tenths

Think about having 16 tenths, that you will distributed into the 4 groups

$$1.6 \div 4 = .4$$

6 groups of _____ hundredths is 2.4

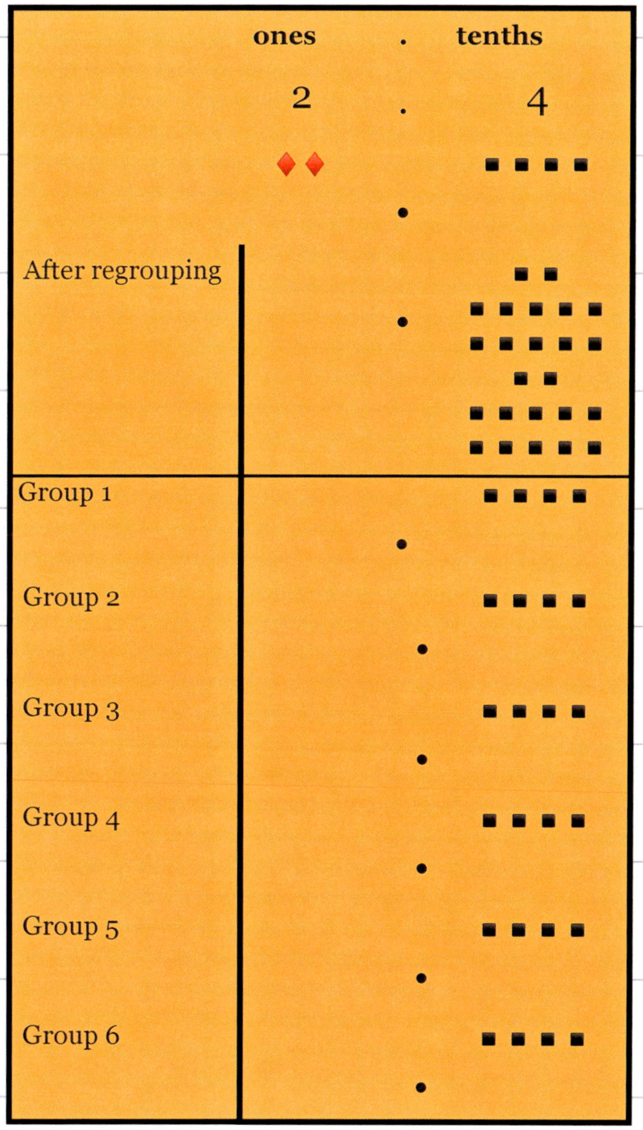

ones	.	tenths
2	.	4

After regrouping

Group 1

Group 2

Group 3

Group 4

Group 5

Group 6

$$2.4 \div 6 = .4$$

Dividing Decimals using Grid Models

seventy five hundredths
divided by three

ex. .75 ÷ 3

seventy five hundredths divided evenly into 3 groups. Each group has .25.

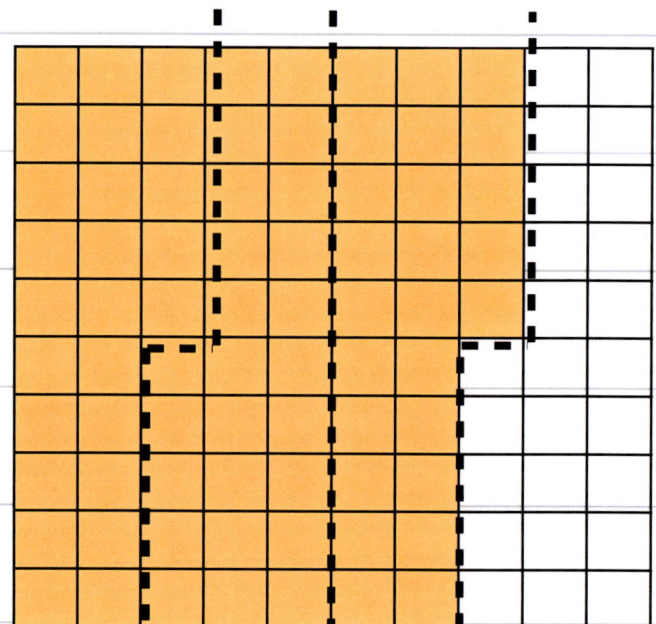

3 groups of twenty-five hundredths

.75 ÷ 3 = .25

seventy five hundredths divided by
three equals twenty five hundredths

Dividing Decimals using Grid Models

one and sixth tenths divided by four

ex. 1.6 ÷ 4

One whole grid and six tenths of the second of a grid is shaded. We make 4 groups. There are 4 tenths in each group.

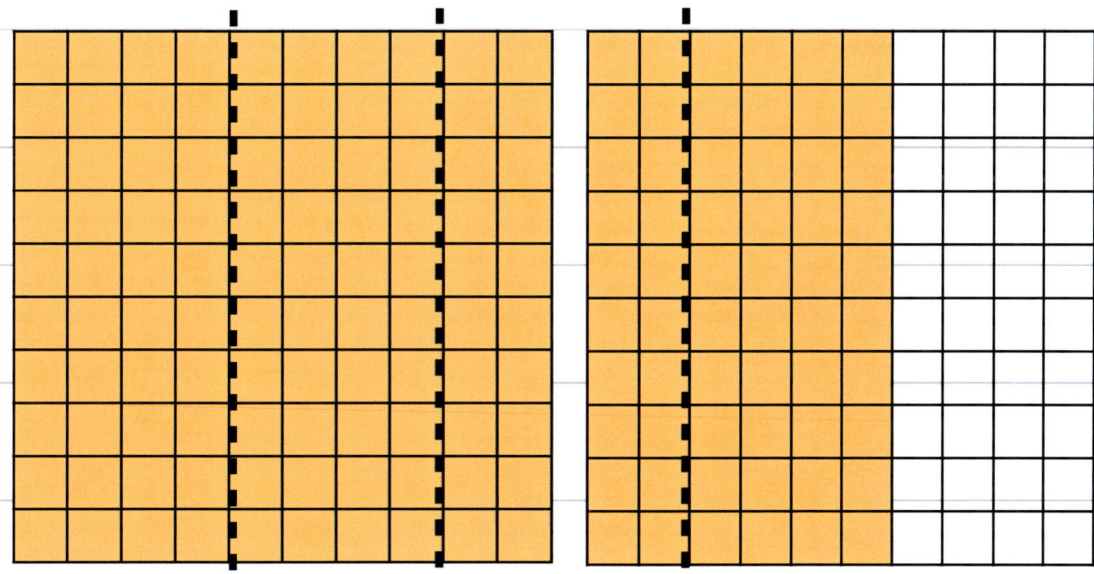

4 groups of 4 tenths

1.6 ÷ 4 = .4
one and sixth tenths divided
by four equals four tenths

Dividing Decimals using Grid Models

two and forty five hundredths
divided by seven

ex. 2.45 ÷ 7

two and forty five hundredths
divided evenly into 7 groups.

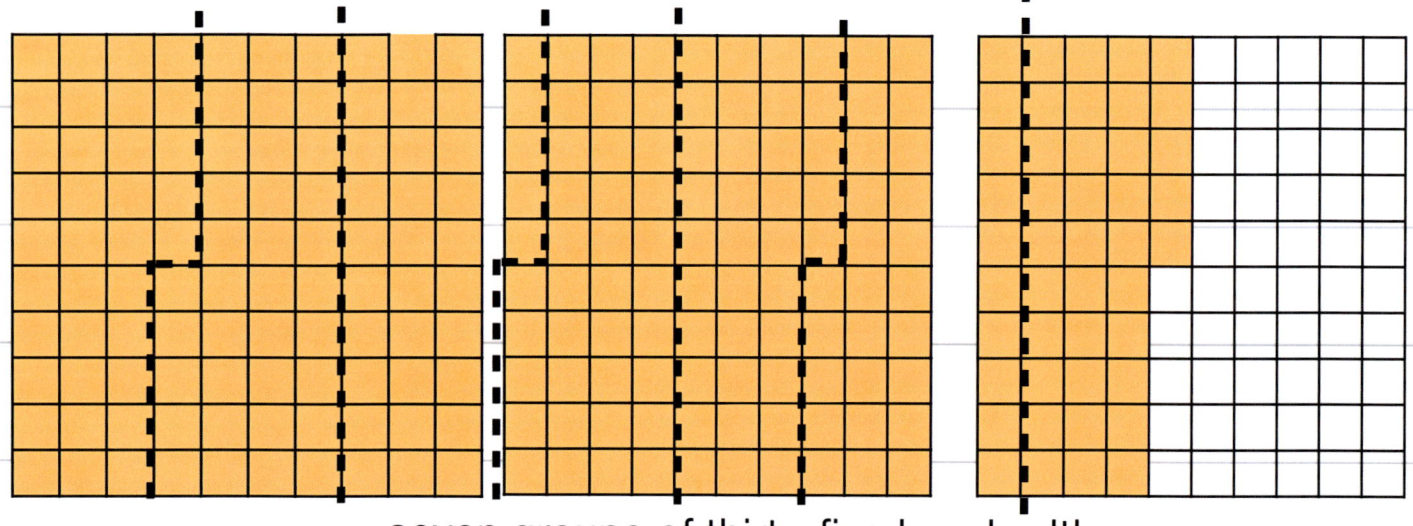

seven groups of thirty-five hundredths

2.45 ÷ 7 = 0.35

Dividing Decimals using Partial Quotients

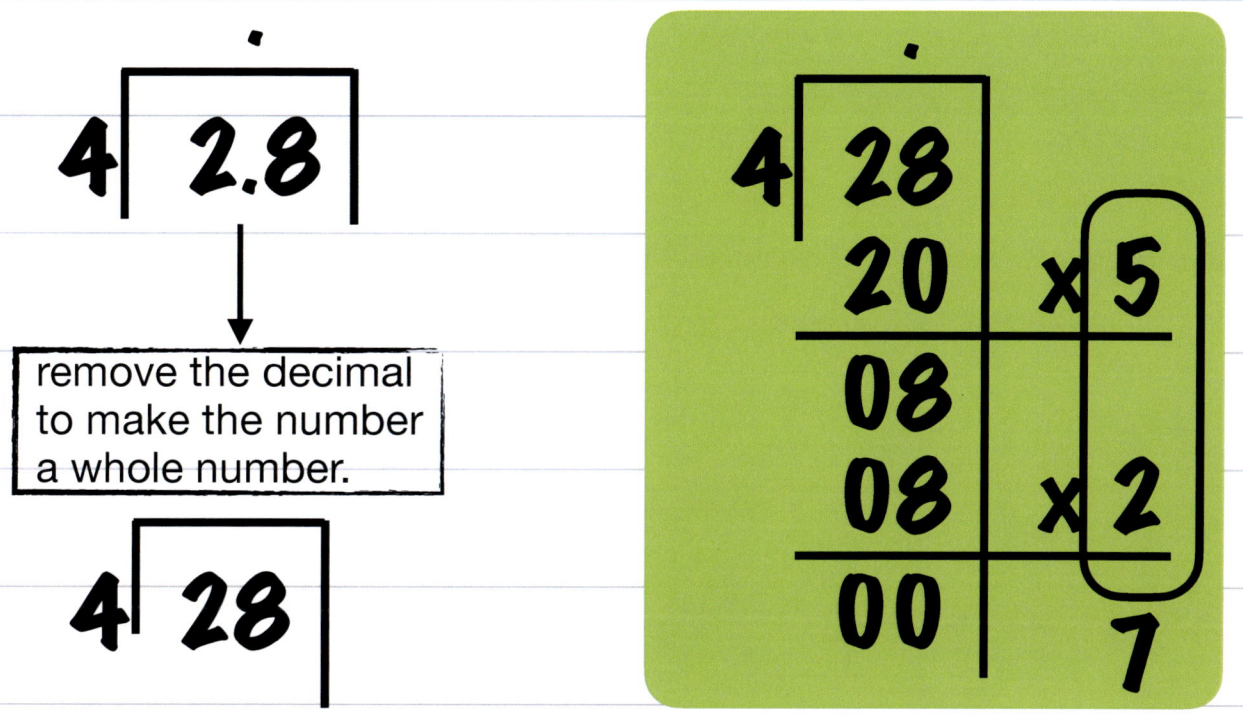

4) 2.8

remove the decimal to make the number a whole number.

4) 28

👣 **1.** Using friendly numbers. Take away at least 5 groups of 4, because **5 x 4 = 20.** We can also take away 2 groups of 4, because **2 x 4 = 8**

👣 **2.** Add the groups, for the quotient.

👣 **3.** Since the dividend has 1 decimal place. The quotient will also have 1 decimal place.

$$2.8 \div 4 = 0.7$$

Dividing Decimals using Partial Quotients

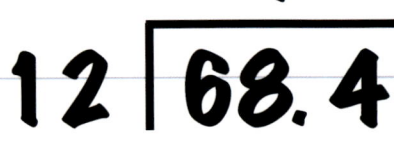

12 | 68.4

> remove the decimal to make the number a whole number.

12 | 68.4

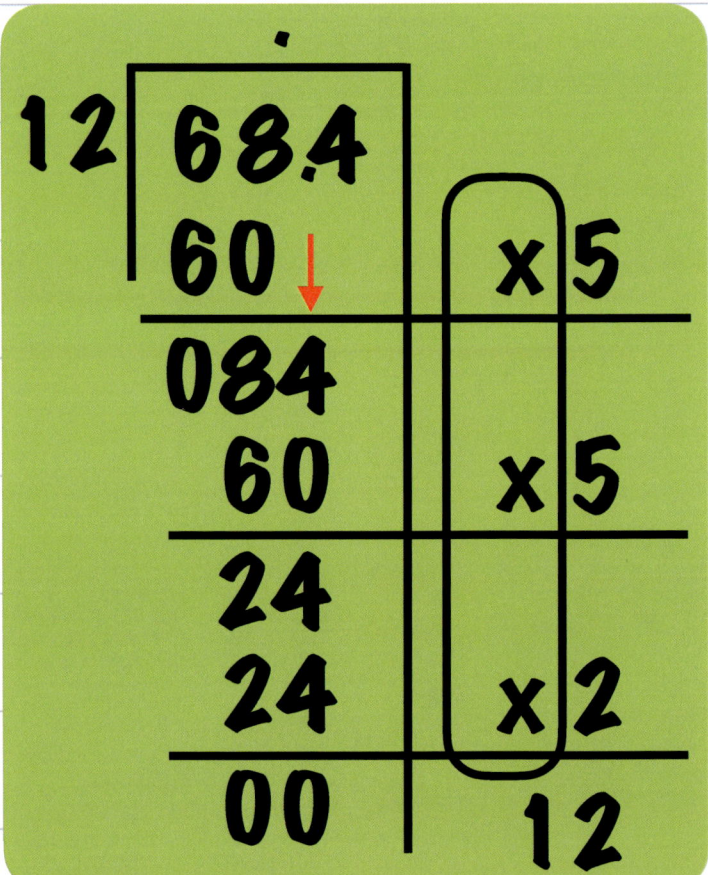

👣 **1.** Using friendly numbers. Take away at least 5 groups of 12, because 5 x 12= 60. We can also take away another 5 groups of 12, because 5 x 12 = 60

👣 **2.** Add the groups, for the quotient.

Since the dividend has 1 decimal place. The quotient will also have 1 decimal place.

$$38.40 \div 12 = 1.20$$

Dividing Decimals using Long Division

ex. 17.45 ÷ 5

seventeen and forty-five hundredths divided by five

There are **5** steps to take when dividing decimals using Long Division method.

1. Divide the first digit of the dividend by the divisor. Write the answer on top as the quotient.

2. Multiply that number with the divisor.

3. Subtract the results from the dividend and write the difference below.

4. Bring down the next digit of the dividend.

5. Repeat until there are none remaining.

Since the dividend has two decimal places. The quotient should have two decimals places

$17.45 ÷ 5 = 3.49$

seventeen and forty-five hundredths divided by five equals three and forty nine hundredths

Division of Decimals

What I Learned

Division of Fractions

Division of Fractions

What I Know	What I want to Know

Dividing Fractions by Whole Numbers

half divided by five

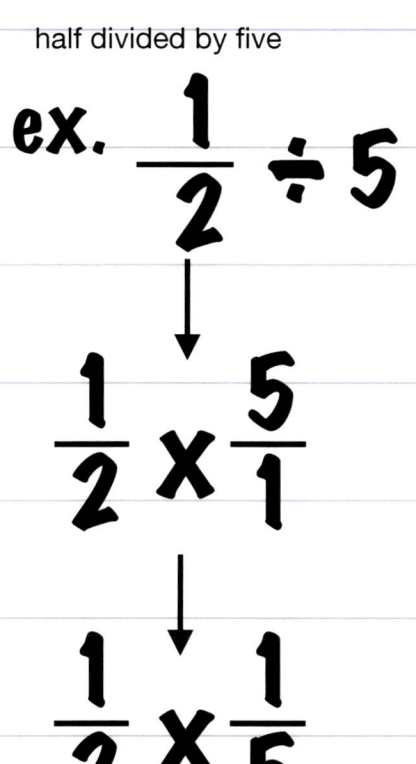

ex. $\dfrac{1}{2} \div 5$

$\dfrac{1}{2} \times \dfrac{5}{1}$

$\dfrac{1}{2} \times \dfrac{1}{5}$

$\dfrac{1}{2} \times \dfrac{1}{5} = \dfrac{1}{10}$

Think about having 1/2 pan of brownies. Sharing the pan of brownies among 5 friends. How much brownie will each person receive?

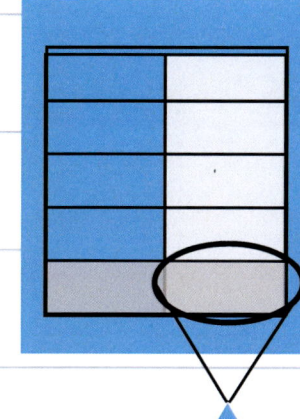

The double shaded portion represents the answer

1. **Keep** the dividend the same

2. **Change** the operation from division to multiplication

3. Make the whole number a fraction by using 1 as the denominator.

4. **Flip**/invert the divisor (by writing the denominator as the numerator and the numerator as the denominator)

5. Multiply the numerators, multiply the denominators.

Dividing Whole Numbers by Fractions

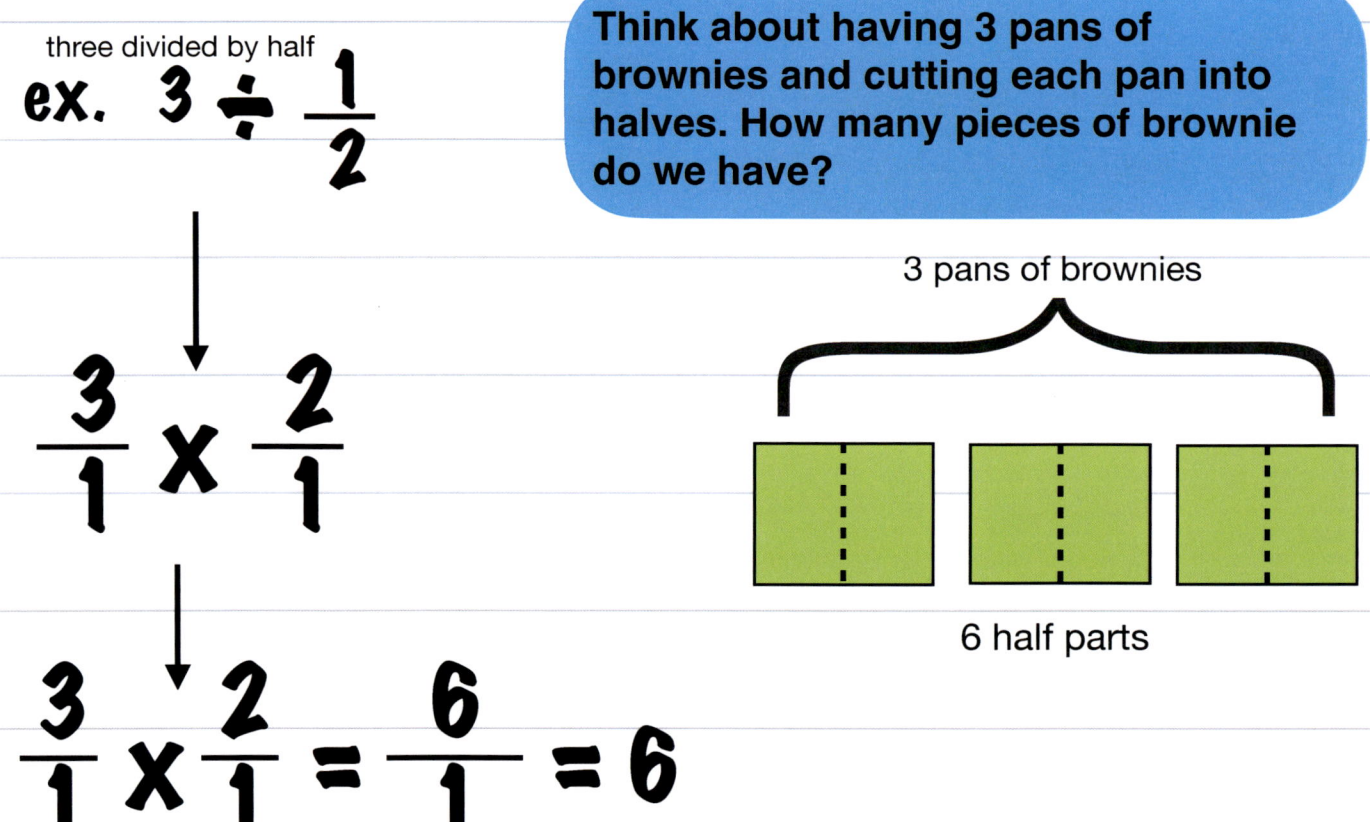

three divided by half

ex. $3 \div \dfrac{1}{2}$

$\dfrac{3}{1} \times \dfrac{2}{1}$

$\dfrac{3}{1} \times \dfrac{2}{1} = \dfrac{6}{1} = 6$

Think about having 3 pans of brownies and cutting each pan into halves. How many pieces of brownie do we have?

3 pans of brownies

6 half parts

👣 1. **Keep** the dividend (whole number) the same . just made into a fraction by using 1 as the denominator.

👣 2. **Change** the operation from division to multiplication

👣 3. **Flip** invert the divisor, by writing the denominator as the numerator and the numerator as the denominator

👣 4. Multiply the numerators, multiply the denominators.

Dividing Fractions by Fractions

Unit fractions by unit fraction

half divided by one forth

$$\frac{1}{2} \div \frac{1}{4}$$

1. **Keep** the dividend the same.

2. **Change** the operation from division to multiplication

3. **Flip**: Invert the divisor, by writing the denominator as the numerator and the numerator as the denominator

$$\frac{1}{2} \times \frac{4}{1}$$

4. Multiply the numerators, multiply the denominators.

5. Convert the improper fraction to a mixed number

$$\frac{1}{2} \times \frac{4}{1} = \frac{4}{2} = 2$$

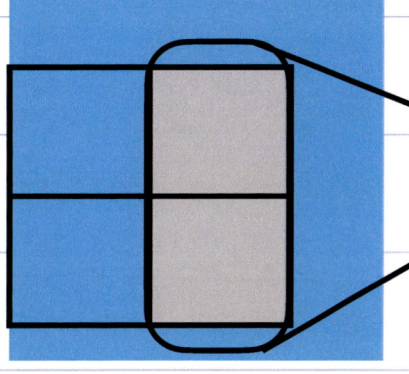

Think about having 1/2 pack of film. Using 1/4 of the pack of film. How much film was used.

Dividing Fractions by Fractions

two thirds divided by three fifths

$$\frac{2}{3} \div \frac{3}{5} \rightarrow \frac{2}{3} \times \frac{5}{3}$$

$$\frac{2}{3} \times \frac{5}{3} = \frac{10}{9} = 1\frac{1}{9}$$

Use division to convert the improper fraction to a mixed number. Think about how many times 9 can go into 10.

1. **Keep** the dividend the same.

2. **Change** the operation from division to multiplication

3. Invert the divisor, by writing the denominator as the numerator and the numerator as the denominator

4. Multiply the numerators, multiply the denominators.

5. Convert the improper fraction to a mixed number

Dividing Mixed Mixed by Fractions

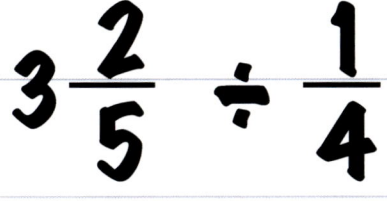

three and two fifths divided by one forth

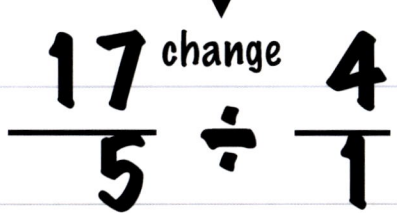

Keep   Flip

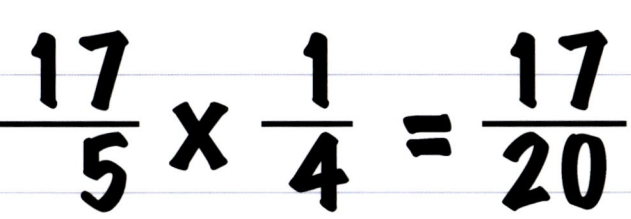

1. Convert the mixed number (dividend) to an improper fraction. Keep the dividend the same

2. Change the operation from division to multiplication

3. Invert the divisor, by writing the denominator as the numerator and the numerator as the denominator

4. Multiply the numerators, multiply the denominators.

5. Convert the improper fraction to a mixed number

Dividing Mixed Numbers by Mixed Numbers

four and three fourths divided by two and one half

$$4\frac{3}{4} \div 2\frac{1}{2}$$

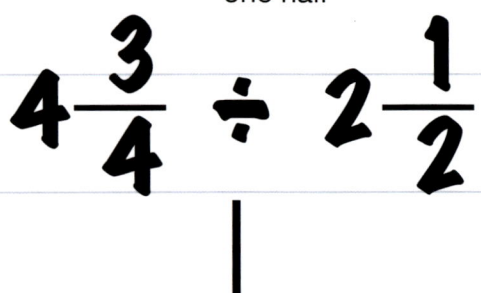

Keep   Flip

1. Convert the mixed numbers (dividend & divisor to improper fractions. Keep the dividend

2. Change the operation from division to multiplication

3. Invert the divisor, by writing the denominator as the numerator and the numerator as the denominator

4. Multiply the numerators, multiply the denominators.

5. Convert the improper fraction to a mixed number

6. Reduce if needed

$$\frac{19}{4} \times \frac{2}{5} = \frac{38}{20} = 1\frac{18}{20} = 1\frac{9}{10}$$

Division of Fractions

Graphic Organizers & Anchor Charts

Anchor Charts

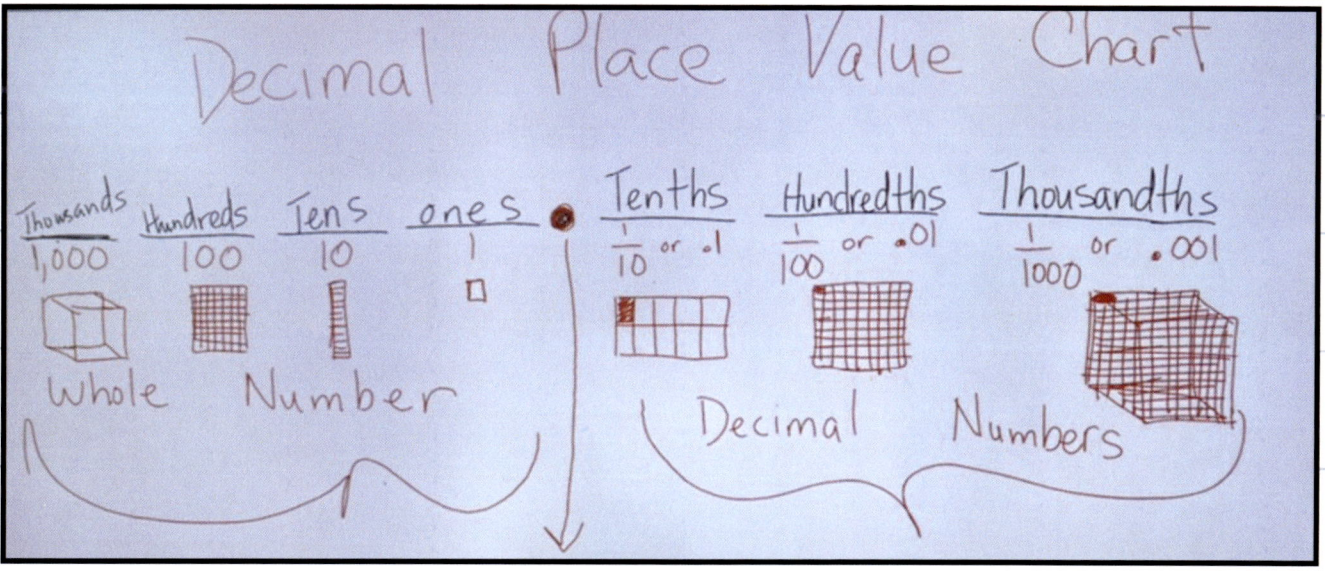

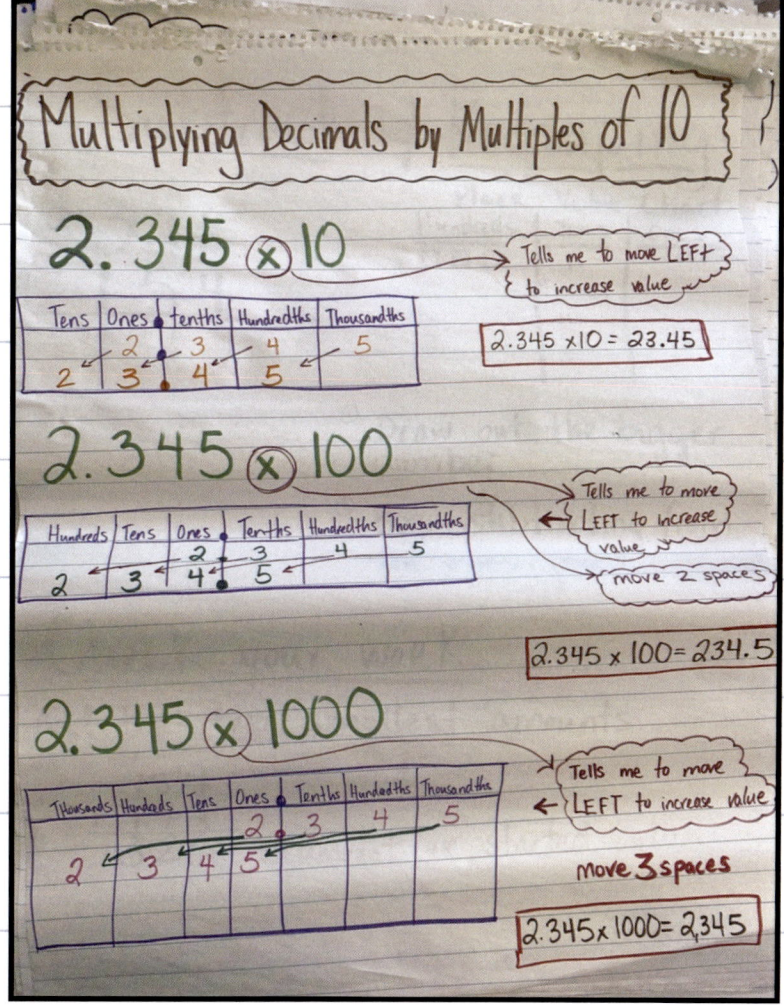

Anchor Charts

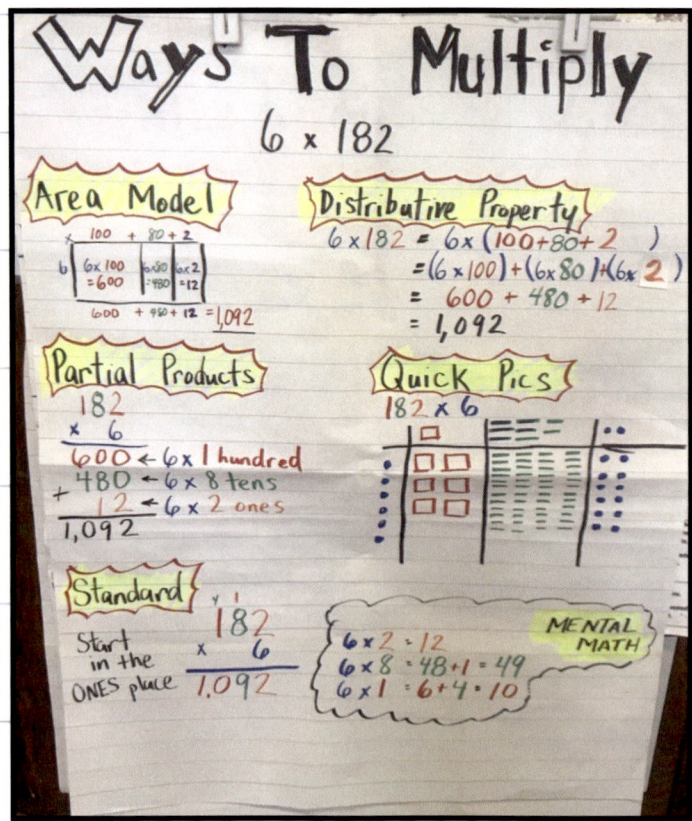

Ways To Multiply
6 × 182

Area Model

	100	+ 80	+ 2
6	6×100 =600	6×80 =480	6×2 =12

600 + 480 + 12 = 1,092

Distributive Property

6 × 182 = 6 × (100+80+2)
= (6×100)+(6×80)+(6×2)
= 600 + 480 + 12
= 1,092

Partial Products

182
× 6
600 ← 6 × 1 hundred
480 ← 6 × 8 tens
+ 12 ← 6 × 2 ones
1,092

Quick Pics

182 × 6

Standard

182
× 6
1,092

Start in the ONES place

MENTAL MATH

6 × 2 = 12
6 × 8 = 48 + 1 = 49
6 × 1 = 6 + 4 = 10

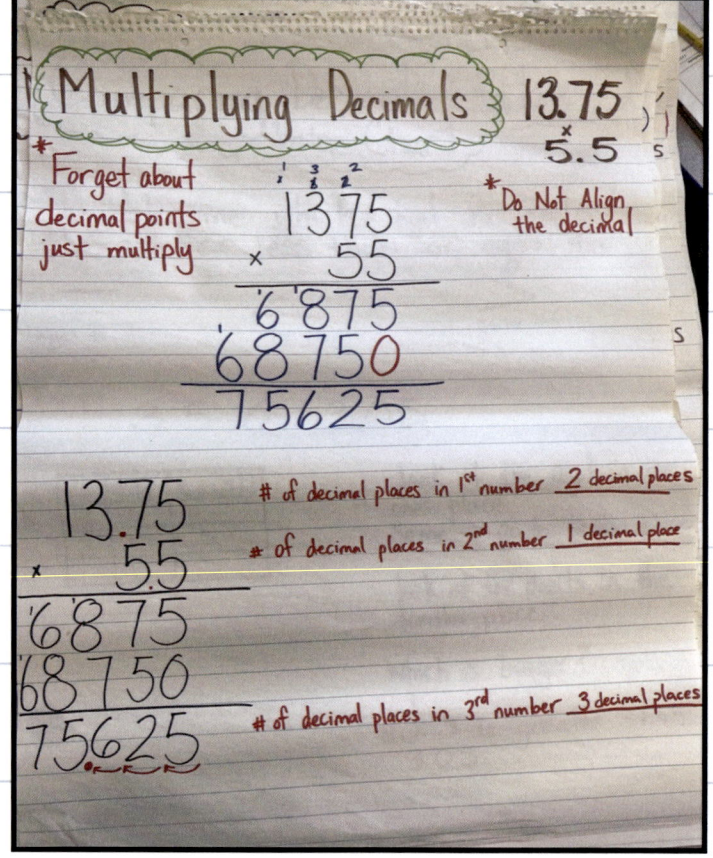

Multiplying Decimals   13.75
                       × 5.5

*Forget about decimal points just multiply

1375
× 55
6875
68750
75625

*Do Not Align the decimal

13.75
× 5.5
6875
68750
75625

# of decimal places in 1st number __2 decimal places__
# of decimal places in 2nd number __1 decimal place__
# of decimal places in 3rd number __3 decimal places__

Anchor Charts

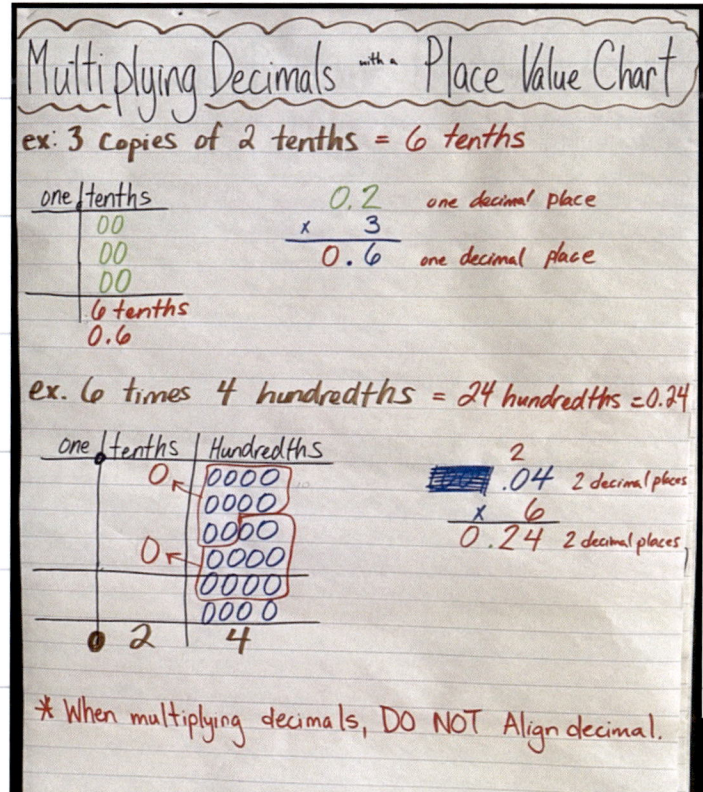

Multiplying Decimals with a Place Value Chart

ex: 3 copies of 2 tenths = 6 tenths

one	tenths
	00
	00
	00

6 tenths
0.6

$$\begin{array}{r} 0.2 \\ \times\ \ \ 3 \\ \hline 0.6 \end{array}$$

one decimal place
one decimal place

ex. 6 times 4 hundredths = 24 hundredths = 0.24

one	tenths	Hundredths
0	0000	
	0000	
	0000	
0	0000	
	0000	
	0000	
0	2	4

$$\begin{array}{r} 2 \\ .04 \\ \times\ \ \ 6 \\ \hline 0.24 \end{array}$$

2 decimal places
2 decimal places

* When multiplying decimals, DO NOT Align decimal.

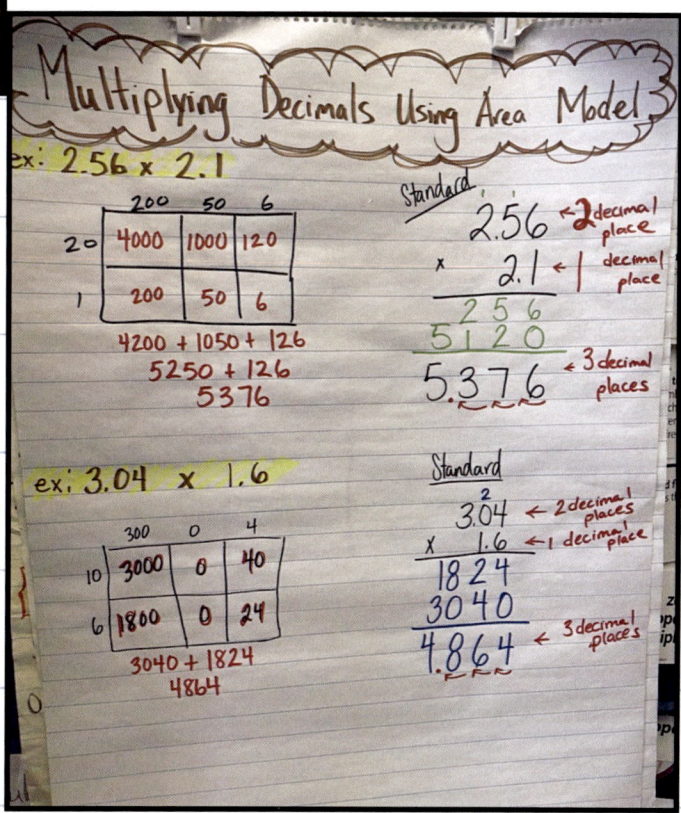

Multiplying Decimals Using Area Model

ex: 2.56 x 2.1

	200	50	6
20	4000	1000	120
1	200	50	6

4200 + 1050 + 126
5250 + 126
5376

Standard

$$\begin{array}{r} 2.56 \\ \times\ \ \ 2.1 \\ \hline 256 \\ 5120 \\ \hline 5.376 \end{array}$$

← 2 decimal place
← 1 decimal place

← 3 decimal places

ex: 3.04 x 1.6

	300	0	4
10	3000	0	40
6	1800	0	24

3040 + 1824
4864

Standard

$$\begin{array}{r} 3.\overset{2}{0}4 \\ \times\ \ \ 1.6 \\ \hline 1824 \\ 3040 \\ \hline 4.864 \end{array}$$

← 2 decimal places
← 1 decimal place

← 3 decimal places

95

Blank Multiplication Charts

Multiplication Tables

Follow this order, complete down and across before moving to the next number

1's 2's 5's 10's 11's

4's 8's 12's

3's 6's 9's

7's

See how fast you can complete your chart!

Graphic Organizer

two by one digit

63 x 7

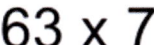

| 60 | + | 3 |

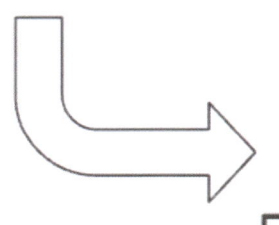

```
7 x 6 =
   42
42 x 10 =
  420
```

```
7 x 3 =
   21
```

```
  420
+  21
_____
  441
```

Step 1: Break apart the first factor into tens and ones. 63 is 6 tens and 3 one or 60 + 3. 7 does not have to be broken apart.

Step 2: Multiply the factors in each column with the factor(s) in the row

Step 3: Add the product from each column 420 + 21

Graphic Organizer

254 x 23

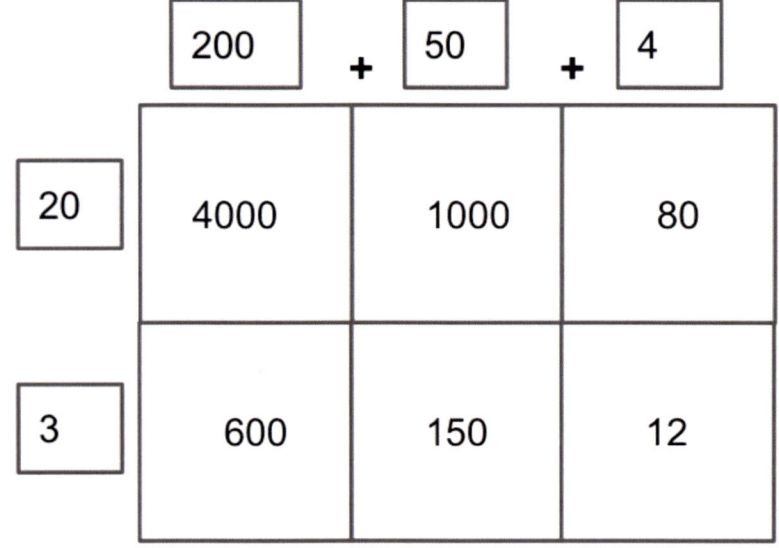

	200 +	50 +	4
20	4000	1000	80
3	600	150	12

Step 1: Break apart the first factor into tens and ones.

Step 2: Multiply the factors in each column with the factor(s) in the row

Step 3: Add the product from each column

$$
\begin{array}{r}
4000 \\
+ \ 600 \\
\hline
4600 \\
+1250 \\
\hline
5850 \\
+ \quad 92 \\
\hline
5942
\end{array}
$$

Graphic Organizer

Blank Area Model - 2 by 2 digit Multiplication

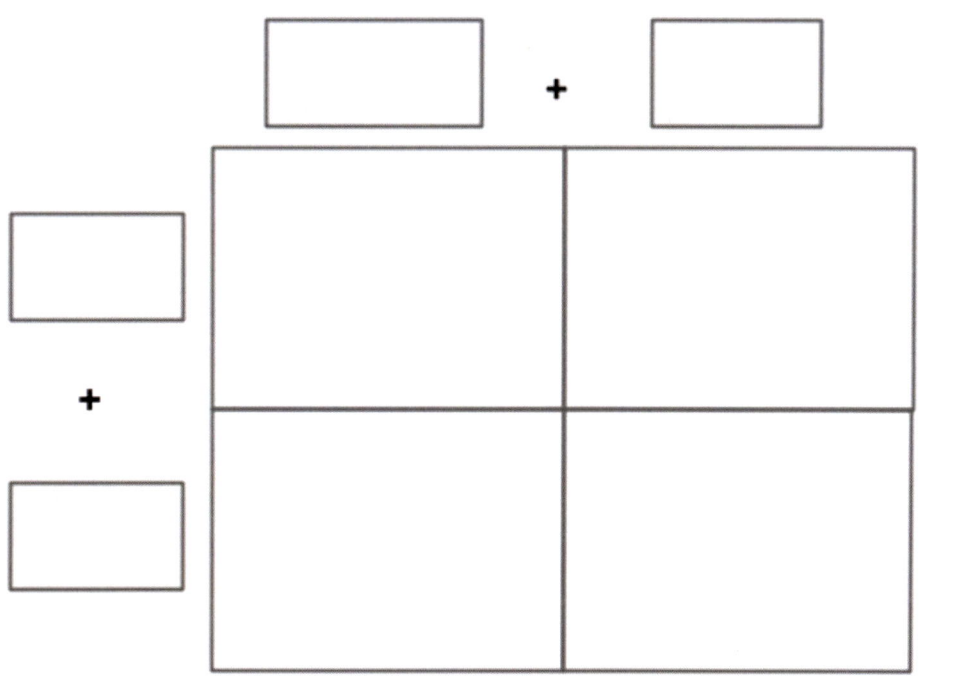

Blank Area Model - 1 by 3 digit Multiplication

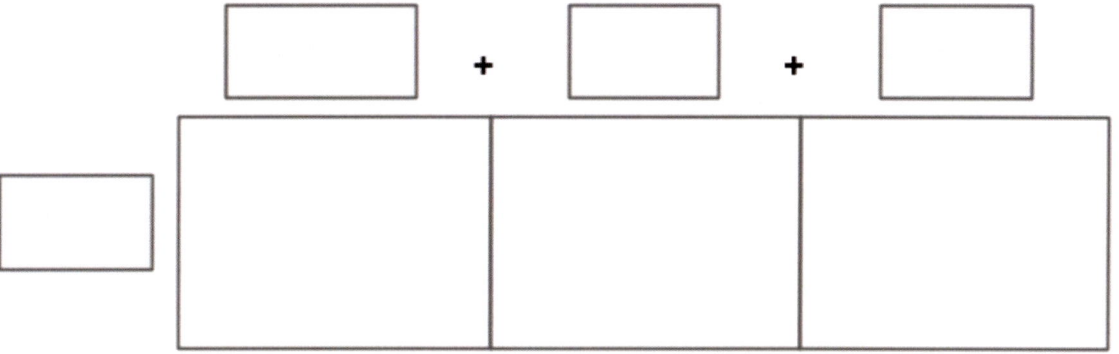

Graphic Organizer

Blank Area Model - 2 by 3 digit Multiplication

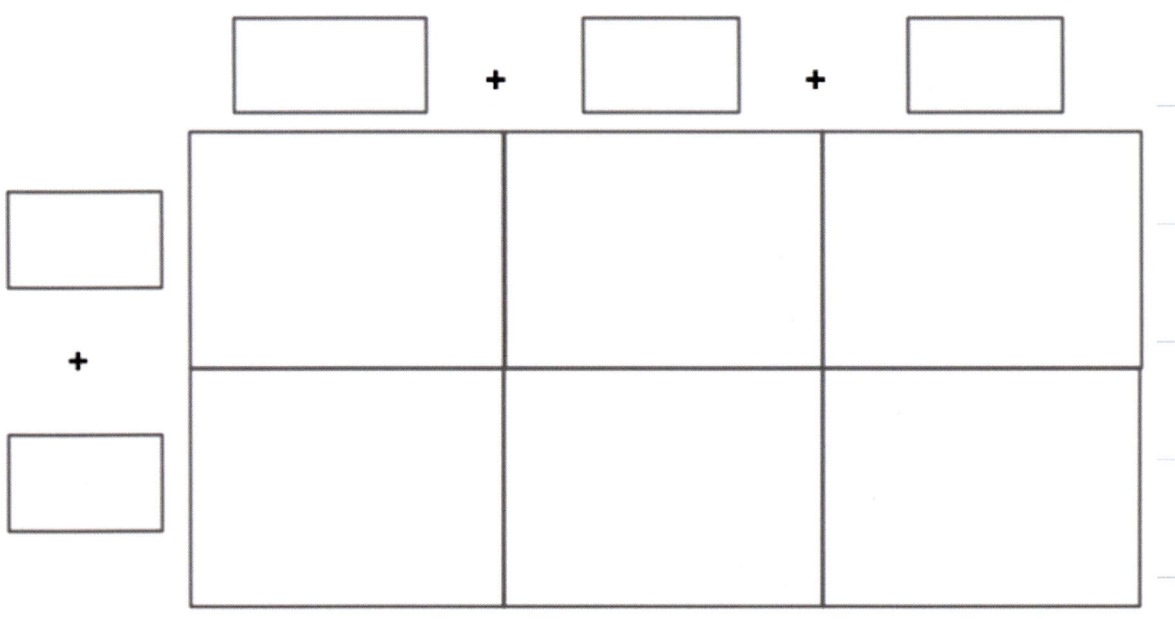

Blank Area Model - 3 by 3 digit Multiplication

Lattice Model Multiplication 2 digit by 2 digit

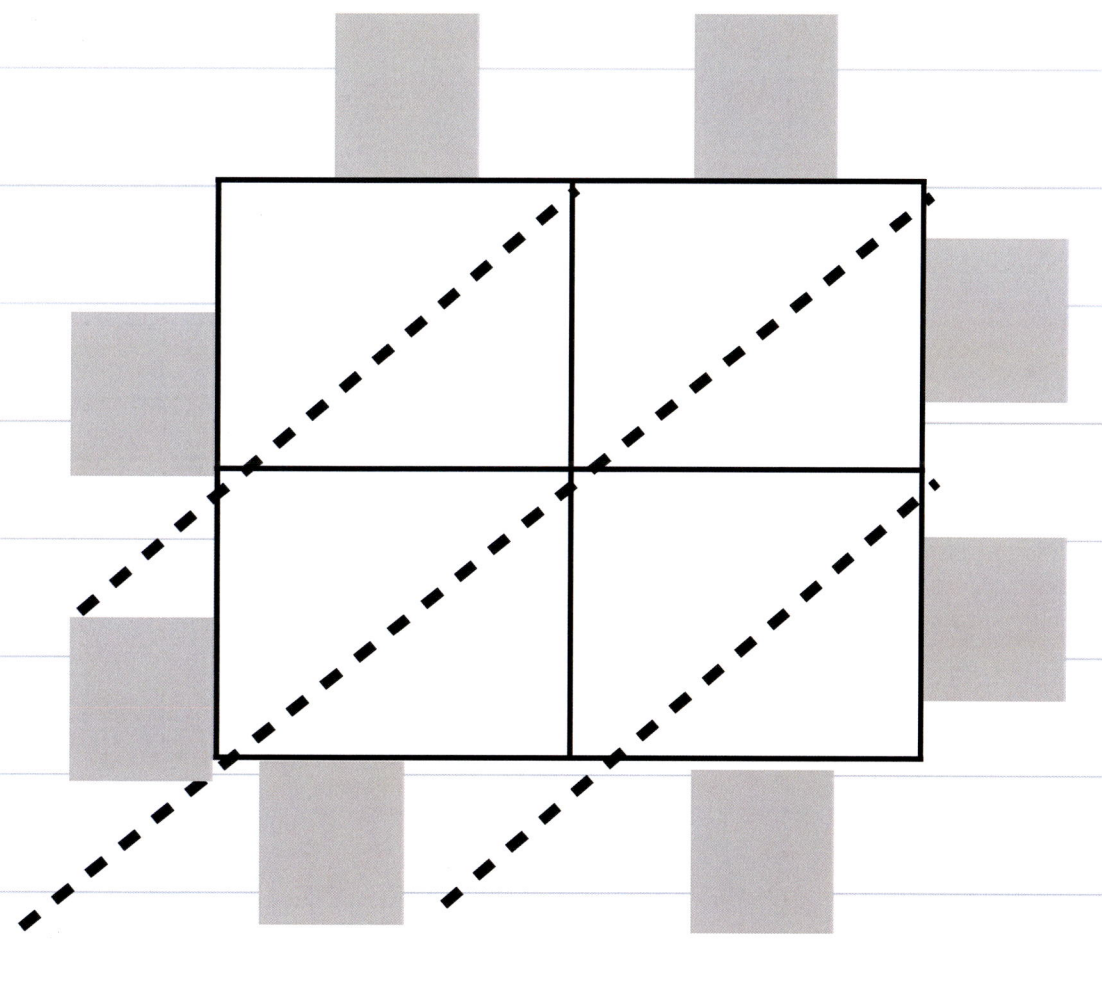

Graphic Organizer

Lattice Model Multiplication 3 digit by 2 digit

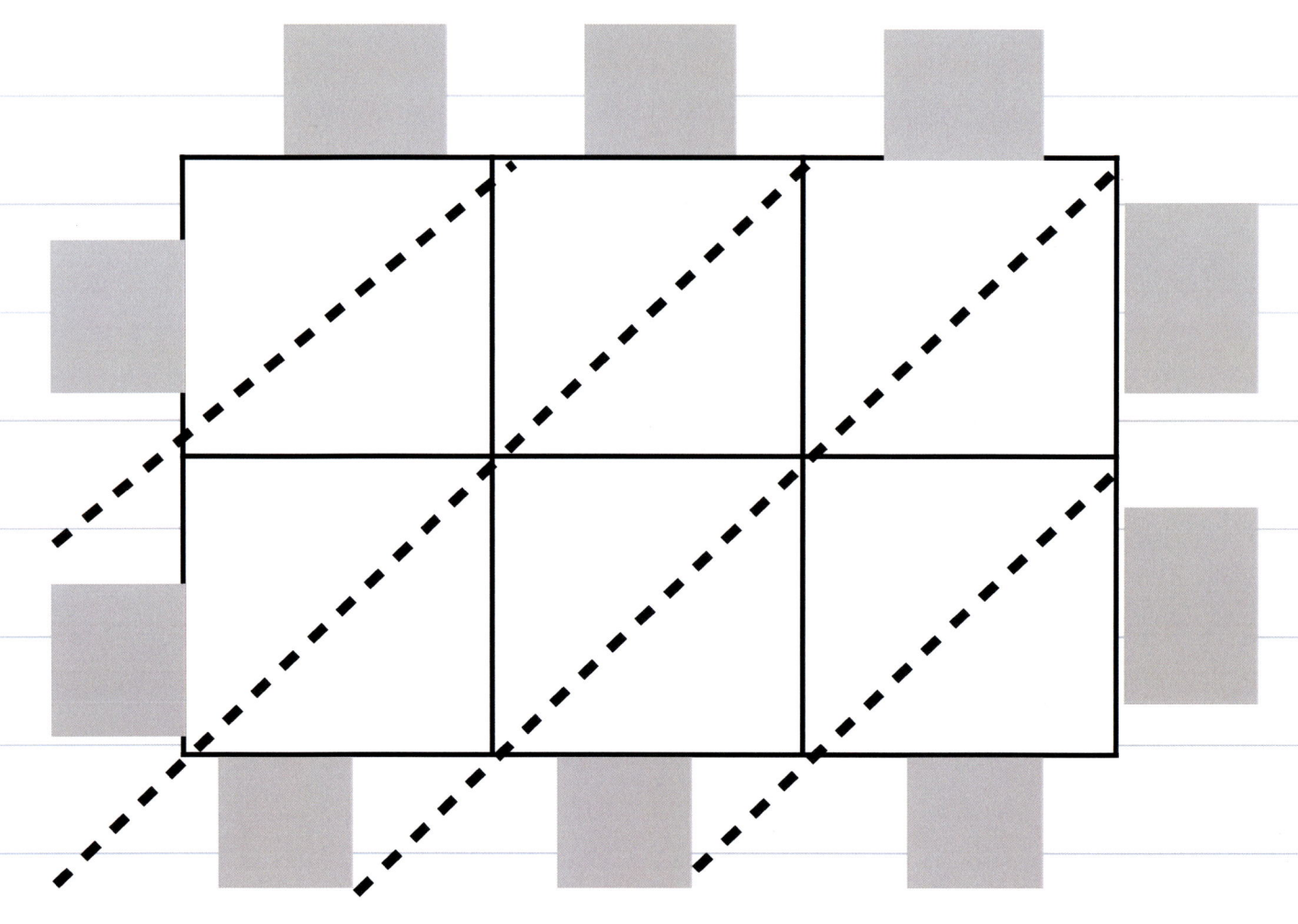

Graphic Organizer

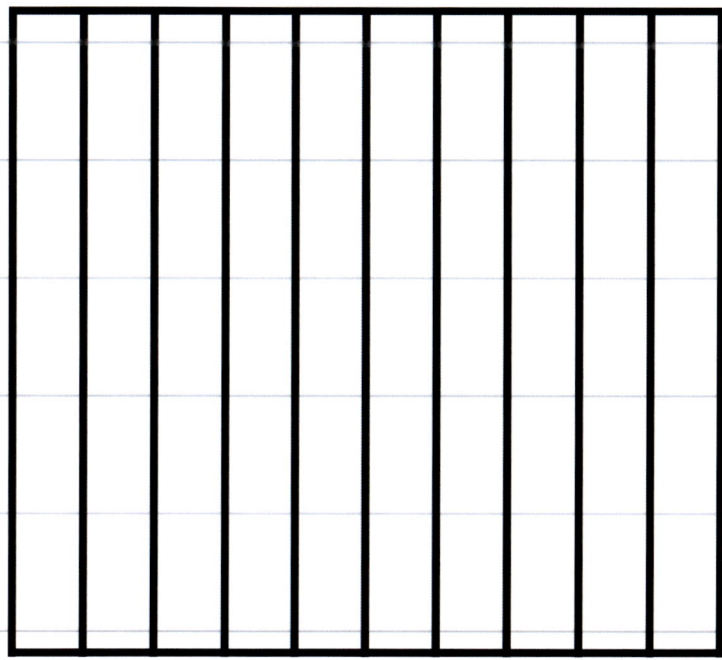

tenths grid model

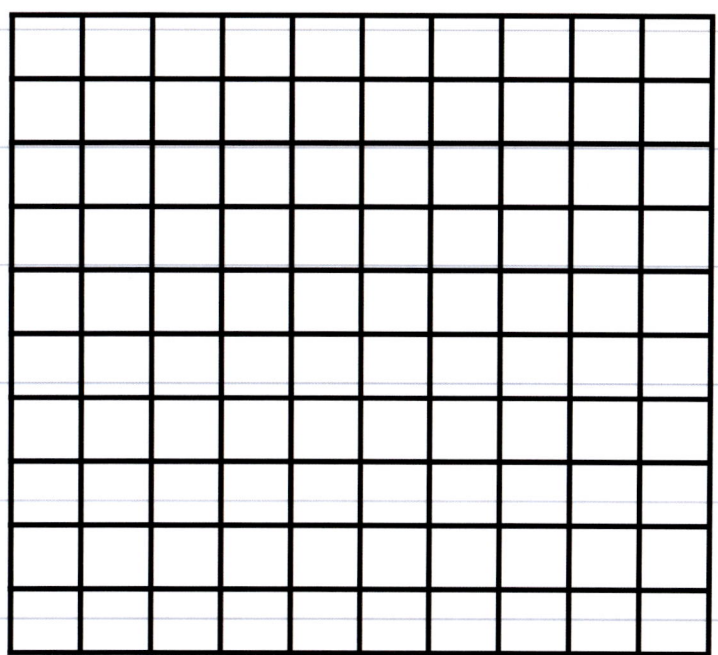

hundredths grid model

Graphic Organizer

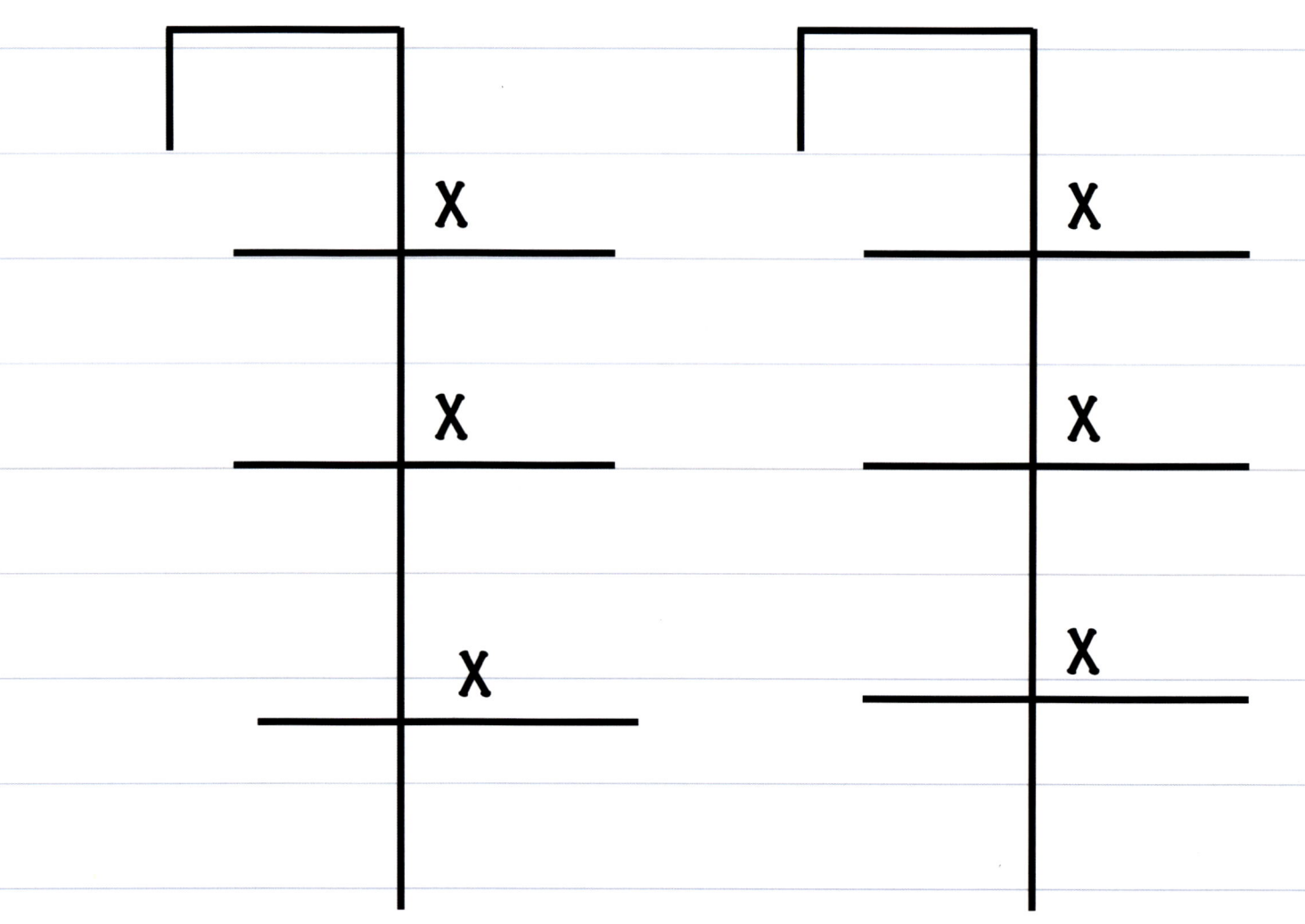

Multiplying Fractions

Created by © Closely Captured LLC

We can multiply a fraction by a whole number

1. Write the whole number as a fraction with a denominator of 1.
2. Multiply the numerators.
3. Multiply the denominators.
4. Simplify, if needed. If your answer is greater than 1, you may want to write your answer as a mixed number.

example

$$\frac{2}{5} \times 3$$

$3 = \frac{3}{1}$ First, write the whole number as a fraction with a denominator of 1.

Now, multiply the numerators and the denominators

$$\frac{2}{5} \times \frac{3}{1} = \frac{2 \times 3}{5 \times 1} = \frac{6}{5}$$

Last, simplify. Since $\frac{6}{5}$ is greater than 1, change it to a mixed number

$$\frac{6}{5} = 1\frac{1}{5} \qquad \text{So, } \frac{2}{5} \times 3 = 1\frac{1}{5}$$

This works because, When you multiply a fraction by a whole number, you can think of the problem as repeated addition.

Looking at the problem ⅖ x 3. that's the same as ⅖ + ⅖ + ⅖

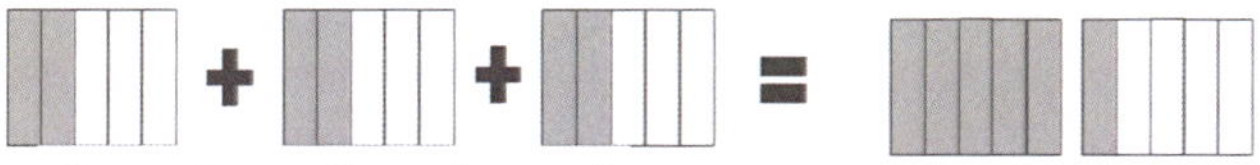

$$\frac{2}{5} \times \frac{3}{1} = \frac{2 \times 3}{5 \times 1} = \frac{6}{5}$$

There are six pieces, and each piece is one fifth of a whole.

Multiplying Fractions

Created by © Closely Captured LLC

Finding the area using fractions

To find the area of a rectangle with fractional sides. Use the formula for area, **length x width.**

Formula = l x w

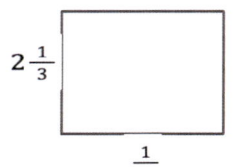

$2\frac{1}{3}$

$\frac{1}{4}$

Mixed Numbers x Fractions

Step 1:→ Change the mixed number to an improper fraction.

<u>*Multiply*</u> *the denominator with the whole number <u>add</u> the numerator over the same denominator*

$$2\frac{1}{3} \ \textbf{\textit{x}} \ \frac{1}{4} \longrightarrow 2\frac{1}{3} = \frac{7}{3} \textbf{\textit{x}} \frac{1}{4} = \frac{7}{12}$$

Refer to notes on converting mixed numbers to improper fractions.

Mixed Number x Whole numbers

Step 1:→ Change the mixed number to an improper fraction.

<u>*Multiply*</u> *the denominator with the whole number <u>add</u> the numerator over the same denominator*

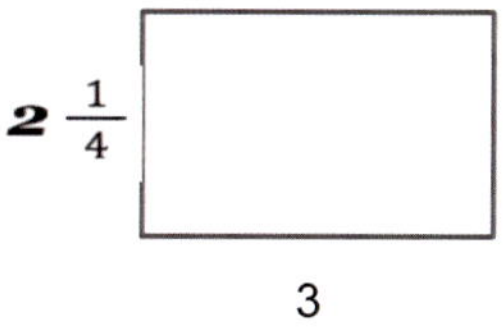

$$\textbf{2}\frac{1}{4} = \frac{9}{4} \textbf{\textit{x}} \frac{3}{1} = \frac{27}{4} = \textbf{6}\frac{3}{4}$$

$\textbf{2}\frac{1}{4}$

3

Mixed Number x Mixed Numbers

Step 1:→ Change the mixed numbers to an improper fractions.

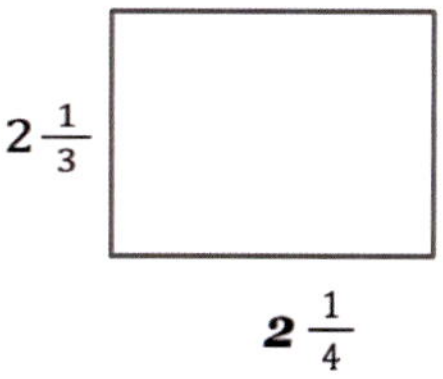

$$\textbf{2}\frac{1}{4} \ \textbf{\textit{x}} \ 2\frac{1}{3} = \frac{9}{4} \textbf{\textit{x}} \frac{7}{3} = \frac{63}{12} = \textbf{5}\frac{3}{12}$$

$2\frac{1}{3}$

$\textbf{2}\frac{1}{4}$

Using Models
Multiplying Fractions

Created by © Closely Captured LLC

1. **Example**. Ms. Borden bakes 4 pans of brownies. She sends ½ the pans to school for her students. How many pans does Ms. Borden send to school?

Model:

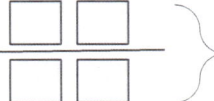

Equation:

½ of 4 =
½ x 4 = 2

4 squares represent the pans of brownies

2. **Example**. Now imagine Ms. Borden has 2 pans of brownies and she sends ½ of the pans to school. How many pans will she send?

Model:

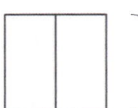

Equation:

½ of 2 =
½ x 2 = 1

2 squares represent the pans of brownies

3. **Example**. Now imagine Ms. Borden has 1 pan of brownies and she sends ½ of the pans to school. How many pans will she send?

Model:

Equation:

½ of 1 =
½ x 1 = ½

1 square represents the pans of brownies

Remember when we multiply fractions we **multiply the numerators, then multiply the denominators.**

4. **Example**. Now imagine Ms. Borden has ½ pans of brownies and she sends ½ of the pans to school. How many pans will she send?

The shaded portion is the ½ pan of brownies

Model:

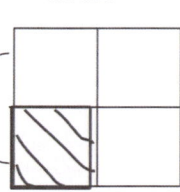

Equation:

½ of ½ =
½ x ½ = ¼

This shaded part represents the portion Ms. Borden sends to school.

5. **Example**. Now imagine Ms. Borden has ¼ pan of brownies and she sends ½ of the pans to school. How many pans will she send?

The shaded area represents ¼ pan of brownies

Model:

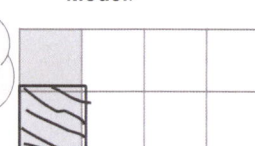

Equation:

½ of ¼ =
½ x ¼ = ⅛

Remember when we multiply fractions we **multiply the numerators, then multiply the denominators.**

Anchor Charts

Multiplying Fractions by Whole Numbers
Multiple Representations

Ms. Borden bought 3 bags of candy. Each bag of candy weighed ¾ of a pound.

Visual model	Equation

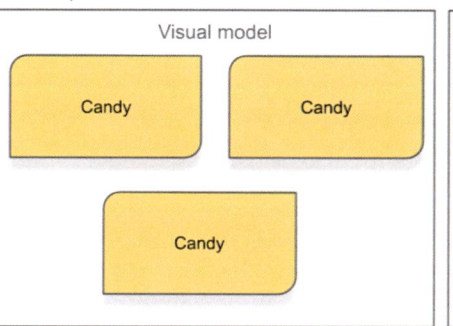

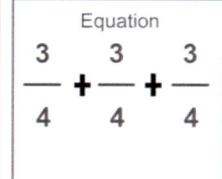

$$\frac{3}{4} + \frac{3}{4} + \frac{3}{4}$$

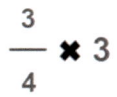

$$\frac{3}{4} \times 3$$

Tape Diagram

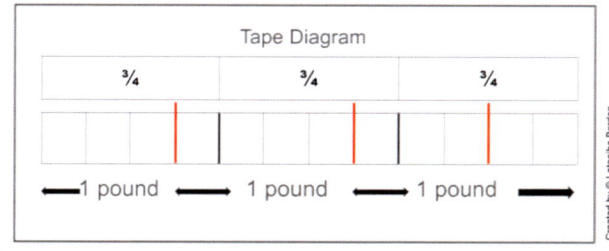

Created by © Lataejha Borden

Numberline

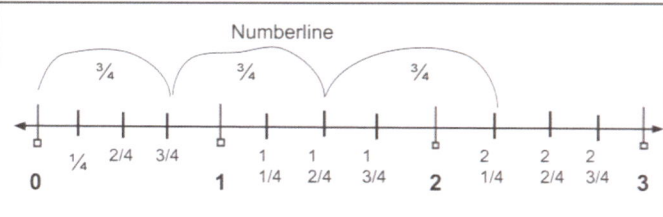

Multiplying Fractions by Whole Numbers

Fractions as a set   using Arrays

$$\frac{1}{7} \text{ of } 14 = 2$$

Make 7 groups. Distribute 14 evenly
among the groups
Select 1 group of 7

$$\frac{3}{7} \text{ of } 14 = 2$$

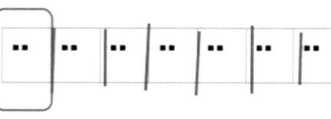

Make 7 groups. Distribute 14 evenly
among the groups
Select 3 group of 7

$$\frac{5}{7} \text{ of } 14 = 10$$

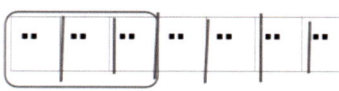
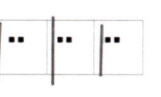

Make 7 groups. Distribute 14 evenly
among the groups
Select 5 group of 7

$$\frac{6}{7} \text{ of } 14 = 12$$

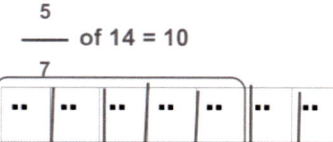
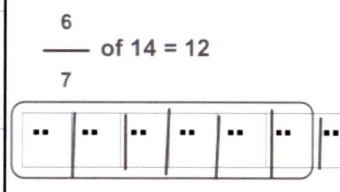
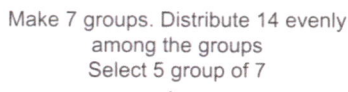

Make 7 groups. Distribute 14 evenly
among the groups
Select 6 group of 7

Created by © Lataejha Borden

107

Anchor Charts

Multiplying Fractions by Whole Numbers

Fractions as a set   using Tape Diagrams

Unit Fractions | Non-Unit Fractions

$\dfrac{1}{2}$ of 4 = **2**

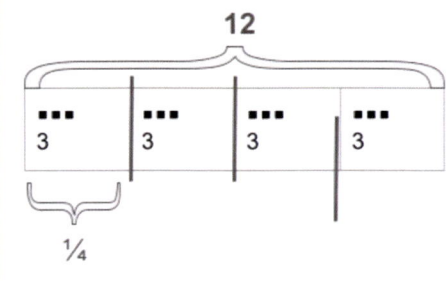

$\dfrac{1}{2}$ of 4 = $\dfrac{1}{2}$ × 4 = $\dfrac{1 \times 4}{2 \times 1}$

$\dfrac{1}{4}$ of 12 = **3**

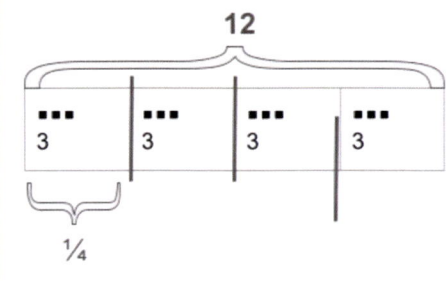

½

$\dfrac{1}{4}$ of 4 = $\dfrac{1}{2}$ × 4 = $\dfrac{1 \times 4}{2 \times 1}$

$\dfrac{2}{3}$ of 9 = **6**

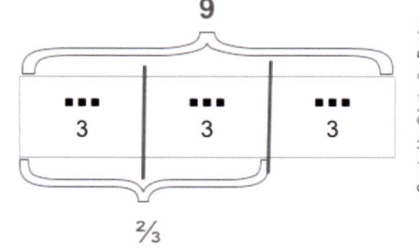

⅔

$\dfrac{2}{3}$ of 9 = $\dfrac{2}{3}$ × 9 = $\dfrac{2 \times 9}{3 \times 1}$

$\dfrac{2}{5}$ of 25 = **10**

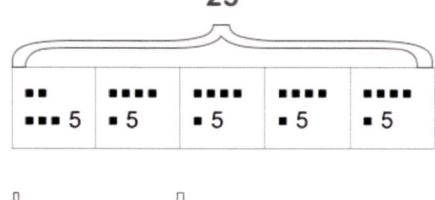

⅖

$\dfrac{2}{5}$ of 25 = $\dfrac{2}{5}$ × 25 = $\dfrac{2 \times 25}{5 \times 25}$

Created by © Lataejha Borden

108

Glossary

Addend/*Addend*	the numbers in an addition problem to be added together to find a sum. *Los números en un problema de adición que se suman para encontrar una suma.*
Array/Orden	an arranged set of rows and columns. Un conjunto organizado de filas y columnas.
Commutative property of Multiplication/ Propiedad conmutativa de la multiplicación	states changing the order of the factors do not change the product. Los estados que cambian el orden de los factores no cambian el producto.
Decimal/Decimal	a number that consists of a whole part and a fractional part, separated by a point. Un número que consiste en una parte entera y una parte fraccional, separadas por un punto.
Denominator/ Denominator	the number at the bottom of the fraction.The number of total parts of the whole or set. El número en la parte inferior de la fracción. El número de partes totales del conjunto o conjunto.
Difference/ Diferencia	the answer to a subtraction problem. La respuesta a un problema de resta.
Digit/Dígito	a number. Un número.
Distributive property of multiplication/ Propiedad distributiva de la multiplicación	states that the sum of two addends multiplied by number gives you the same answer. Afirma que la suma de dos agregas multiplicada por el número te da la misma respuesta.
Divide/Dígito	the process sharing out equally. El proceso se comparte por igual.

Dividend/ Dividendo	a number to be divided by another number. Un número que se dividirá por otro número.
Equal/Igual	to have the same value in, size, amount or quantity. Tener el mismo valor, tamaño, cantidad o cantidad.
Equal groups/ Grupos iguales	groups with the same number of objects. Grupos con el mismo número de objetos.
Equation/ Ecuación	a number sentence with an equal sign. Everything on one side of the equal sign has to be equal to the other side. Una oración numérica con un signo igual. Todo lo que hay en un lado del signo igual tiene que ser igual al otro lado.
Equivalent/ Correspondiente	equal in amount or value. Igual en cantidad o valor.
Equivalent fraction/ Fracción equivalente	fractions that are equal in value but look differently, have different dominators. Fracciones que son iguales en valor pero tienen un aspecto diferente, tienen diferentes dominantes.
Factor(s)/ Factor(s)	a number(s) that multiply together to make a product. Un(s) número(s) que se multiplican para hacer un producto.
Fluency/ Fluidez	the ability the perform mathematical problem solving effective. La capacidad de realizar la resolución de problemas matemáticos de manera efectiva.
Fraction/ Fracción	number smaller than a whole number. Número menor que un número entero.
Horizontal/ Horizontal	parallel to the plane of the horizon. Paralelo al plano del horizonte.
Identity Property/ Propiedad de identidad	states that any number multiplied by 1 will be that number. Any number divided by 1 will be that number. Establece que cualquier número multiplicado por 1 será ese número. Cualquier número dividido por 1 será ese número.

Improper fraction/ Fracción incorrecta	a fraction that has a numerator greater than the denominator. Una fracción que tiene un numerador mayor que el denominador.
Invert/Invert	reverse, turn upside down. Invierte, gira boca abajo.
Inverse/ Opposite	opposite operations. addition is inverse to subtraction and multiplication is inverse to division. Operaciones opuestas. La suma es inversa a la resta y la multiplicación es inversa a la división
Long division/ División larga	the method of breaking apart larger numbers into smaller groups. El método de dividir números más grandes en grupos más pequeños
Multiple/ Múltiple/	a number that can be divided by another number. Un número que se puede dividir por otro número.
Multiplicand/ Multiplicando	a quantity which is to be multiplied by another (the multiplier). Una cantidad que se va a multiplicar por otra (el multiplicador).
Multiplier/ Multiplicador	a quantity by which a given number (the multiplicand) is to be multiplied. Una cantidad por la que se va a multiplicar un número dado (el multiplicando).
Non unit fraction/ Fracción no unitaria	a fraction that has a numerator that is not 1. Una fracción que tiene un numerador que no es 1.
Number line/ Línea numérica	a line with a set of ordered numbers, the numbers are represented by tick marks. Una línea con un conjunto de números ordenados, los números están representados por marcas de verificación.
Numerator/ Numerator	the number at the top of the fraction. The number of parts considered or selected. El número en la parte superior de la fracción. El número de piezas consideradas o seleccionadas.
Partition/ Partición de disco	to divide into equal parts. Dividir en partes iguales.

Place value/ Valor de lugar	the value of a space given to a number. The spaces named according to a value. El valor de un espacio dado a un número. Los espacios nombrados según un valor.
Place value system/ Sistema de valor de lugar	the system in which the position of a digit determines that value. El sistema en el que la posición de un dígito determina ese valor.
Product/ Producto	the answer to a multiplication problem. La respuesta a un problema de multiplicación.
Proper fraction/ Fracción propia	a fraction that has a numerator smaller than the denominator. Una fracción que tiene un numerador más pequeño que el denominador.
Quotient/ Quotient	the answer to a division problem. La respuesta a un problema de división.
Reciprocal/ Recíproco	the inverse of a number or value. La inversa de un número o valor.
Reduce/ Reducir	to find the smallest factor, also known as simplifying and finding the lowest terms. Para encontrar el factor más pequeño, también conocido como simplificar y encontrar los términos más bajos.
Remainder/ Restante	the number that is leftover after the process of division is done. The remainder is smaller than the divisor. El número que queda después de que se complete el proceso de división. El resto es más pequeño que el divisor.
Symbols/ Símbolos	Letter's or marks that represent a quantity, operation or relationship. Letras o marcas que representan una cantidad, operación o relación.
Unit fraction/ Unidad de fracción	a fraction with 1 as the numerator. Una fracción con 1 como numerador.

Vertical/ Vertical	goes up and down. Sube y baja.
Whole number/ Número entero	a number without a decimal or fraction part. Un número sin una parte decimal o fraccional.

About the Author

Lataejha Borden has been an elementary school teacher for more than 10 years with the New York City Department of Education. After graduating with an Associate degree in Computer Science, she went on to earn a bachelor's degree from St. Francis College. She earned her Master's Degree in dual General and Special Education at Touro College.

Lataejha's love of working with students, her understanding that students need additional reinforcements, and parents need up to date guides to help their children. Knowing there is a need for resources to help elementary and middle school students, and the people that support them. Lataejha's own love for learning and helping the community to enjoy and understand mathematical concepts led to the development of this book.

This book helps the struggling students and their supporters. This book will provide supports that will help find and maintain a love for math.

www.ingramcontent.com/pod-product-compliance
Lightning Source LLC
Chambersburg PA
CBRC090843120626
46551CB00009B/740